THINKING
ABOUT
SCIENCE

MAX DELBRÜCK AND
THE ORIGINS OF
MOLECULAR BIOLOGY

THINKING
ABOUT
SCIENCE

*MAX DELBRÜCK AND
THE ORIGINS OF
MOLECULAR BIOLOGY*

Ernst Peter Fischer and Carol Lipson

W. W. Norton & Company · New York · London

The text of this book is composed in Aster, with display type set in Galliard. Composition and manufacturing by The Haddon Craftsmen, Inc. Book design by Bernard
Klein.

Library of Congress Cataloging-in-Publication Data
Fischer, Ernst Peter, 1947–
    Thinking about science : Max Delbrück and the origins of molecular biology /
Ernst Peter Fischer, Carol Lipson.
        p. cm.
    Includes index.

    1. Delbrück, Max. 2. Molecular biologists—United States-
-Biography. I. Lipson, Carol. II. Title.
QH31.D434F573 1988
574.8′8′0924—dc19
        [B]   87-16626

ISBN 0-393-02508-X
ISBN 0-393-96084-6 (pbk.)

W. W. Norton & Company, Inc., 500 Fifth Avenue, New York, N. Y. 10110
W. W. Norton & Company Ltd., 37 Great Russell Street, London WC1B 3NU

    2 3 4 5 6 7 8 9 0

Contents

Preface

The Pleasure of Thinking About Science

This book was written for friends of science. It describes the life of Max Delbrück, who began in the oldest and started the youngest science. Max's scientific activities span a surprisingly wide range. He studied astronomy, received a Ph.D. in theoretical physics, contributed to the understanding of the atomic nucleus, applied quantum mechanics to genetics, was instrumental in creating molecular biology, turned to the analysis of sensory behavior, and sought epistemological insights from science. Beginning as a student of astronomy, Max became at the end a student of evolution.

In 1969, Max shared the Nobel Prize for Physiology and Medicine with Salvadore Luria and Alfred Hershey. Since they were the first to publish on bacterial genetics, their work opened the field. Yet even as the object of the highest recognition possible in the scientific world, Max remained totally unpretentious. He always directed friends and coworkers to call him Max; anyone referring to him as Dr. Delbrück or Professor Delbrück would have been rebuked. This biography, then, will refer to him as he wished to be addressed.

Max was the intellectual of molecular biology. There are others whose original contributions had greater impact on the development of science, but it was Max whose nagging questions guided researchers toward the right experiments. He was the Socrates of biology; Cold Spring Harbor and Caltech were his marketplaces.

Those who worked with him will forever remember his reaction to any discovery they felt ready to announce: "I don't believe a word of it." Discussing experiments with Max meant receiving a preview of all possible criticism to come. He was hard to convince, yet willing to spend much time in mastering details. Despite the eternal skepticism, no one could fail to note how much pleasure Max derived from thinking about science. Nor could one fail to admire his openness, his genuine humility, and his scientific integrity. Max's presence enforced quality; even jokes had to be good ones. There were always a lot of jokes around Max.

EN ROUTE TO A BIOGRAPHY

The history of this book dates back to 1978, when Max was interviewed by Carolyn Kopp. A graduate student at UCLA, she had planned to write her dissertation in science history on Max's life. Max agreed to cooperate. The carefully edited interviews are now deposited in the archives at Caltech. Since Max's career spans many scientific areas and two languages, and since Carolyn's training was as a science historian rather than as a scientist, she ultimately concluded she was not the right person for the job.

Through conversations with Carolyn, Max came to realize that many young science historians were struggling to obtain information about people he had known, who had failed to provide their own accounts. Max became aware of the difficulty historians face in reconstructing the circumstances of their subjects' lives. He decided to do the job himself, describing it as "an entirely new attempt to come to grips with the problem of capturing some kind of inner unity in my life." Max died shortly after dictating this sentence. A victim of cancer, he knew how little time remained when he set out on the autobiographical project. He asked one of us (Peter) to come to Pasadena to help him with the book.

Peter had been one of the last graduate students in Max's laboratory at Caltech, having left in 1977. In 1981, he was a researcher in biophysics at the University of Konstanz, with two awards for science writing in German. In February of that year, Peter joined Max in Pasadena to discuss the biography; they agreed on the description of Max's life in three parts with four chapters each to which this book remains faithful. Together, they developed headings for each unit. The preliminary work encouraged Max to begin dictating his memoirs. He managed about thirty-five pages of notes before his death in March 1981. While carrying on with the task of writing in a non-native language, Peter asked the second author, Carol, for help.

To a great extent, we have used Max Delbrück's own words in this study of his contributions to science. Quotes from Carolyn Kopp's interviews with Max are referred to as (I). Quotes from Max's publications, which are listed in the Bibliography, are numbered D 1 through D 111 for purposes of reference. Material quoted from the autobiographical pages Max dictated is designated (A). We have also used Max's unpublished letters and papers; these are deposited in the archives at Caltech.

Max spent all of his life enthusiastically learning or doing science. But he was well aware that science "does not solve our problems" (I). He appreciated the limits of the scientific method, which is incapable of addressing fundamental questions of human existence.

Neither did Max devalue science. In his view, science only finds its proper home in society if that society possesses some understanding of what scientists do, and of what they cannot do. The scientific story this book tells is meant to aid in such a goal. It illustrates the adventure that science provides, the pleasure one can derive from it, and the struggles a scientist faces.

The book describes the life of a man who, more than any other biologist, has shown how the power of personality can influence the progress of science. Max derived this power mainly through his personal probity and intellectual scrupulousness. He regularly avoided assuming positions of administrative or political power; in his view, such positions would only have distracted him from his search for truth and from the pleasures of science.

ACKNOWLEDGMENTS

This project would have been difficult at best and impossible at the worst without the help and support of many individuals and institutions. The Deutsche Forschungsgemeinschaft granted Peter a two-year fellowship, as well as providing travel funds. The Syracuse University Research Fund provided summer support to allow the two of us to meet in Germany to collaborate. In Konstanz, Dr. Horst Sund, president of the university, was instrumental in helping solve a variety of problems, not least of which involved office space and typing facilities. Hildegard Allen was a great help with the typing.

To the Delbrücks, including Max's sister Emmi Bonhoeffer, we are deeply grateful. Max's wife Manny and the children have included us as members of the family. All have been generous with their reminiscences. Manny has supported the work financially, by supplying travel funds and a word processor. Beyond that, she has offered invaluable moral support and advice. Her friendship has been one of the greatest rewards of this enterprise.

Max's former colleagues have also generously opened their doors. In particular, James Watson hosted Peter on the grounds of Cold Spring Harbor Laboratory, which had served as Max's second home for years. Gunther Stent provided crucial support and counsel. Many others have spoken with us and provided help. Though we would wish to describe the contributions of each, we must limit ourselves to citing their names in gratitude: Finn Aaserud, Copenhagen; Gerold Adam, Konstanz; Seymour Benzer, Pasadena; Kostia Bergman, Boston; Ruth and Felix Boehm, Pasadena; Aage Bohr, Copenhagen; Carsten Bresch, Freiburg; Beate Carrière, Pasadena;

Enrique Cerdá-Olmedo, Seville; Marga Dütting, West Berlin; Wolfgang Eckardt, East Berlin; Arturo Perez Eslava and Maribel Alvarez, Salamanca; Doris and Caleb Finch, Pasadena; Adolf Henning Frucht, West Berlin; Erhardt Geissler, East Berlin; Janos Hajdu, Cologne; Martin Heisenberg, Würzburg; Alfred Hershey, Cold Spring Harbor, New York; George P. Hess, Ithaca; Anita Hoffmann, Bonn; Norman Horowitz, Pasadena; Lothar Klünner, West Berlin; Carolyn Kopp, New York City; Peter Läuger, Konstanz; Edward D. Lipson, Syracuse; Salvadore Luria, Cambridge, Massachusetts; Stuart McLaughlin, New York City; Karl von Meyenn, Stuttgart; Bernd Mühlschlegel, Cologne; Tamotsu Ootaki, Yamagata; Manfred and Ursel Otto, Tübingen; David Presti, Eugene, Oregon; Werner Reichhardt, Tübingen; Enzo Russo, West Berlin; Peter Schuster, West Berlin; Peter Starlinger, Cologne; Josef Straub, Cologne; Thomas Trautner, West Berlin; Bob Walker, Tesuque, New Mexico; Victor Weisskopf, Cambridge, Massachusetts; Carl Friedrich von Weizsäcker, Starnberg; Tazewell Wilson, New York City; and Yno and Lilly Yan, San Francisco.

THINKING
ABOUT
SCIENCE

MAX DELBRÜCK AND
THE ORIGINS OF
MOLECULAR BIOLOGY

New Physics in the Old World

"I developed the habit of not taking any notes at lectures but watching the professor and trying to follow the mathematical argument with some understanding. When the professor made a little mistake, with a plus or a minus sign or a factor 2, I did not point that out directly but waited ten minutes until he got entangled and then pointed out, to his great relief, how he could disentangle himself—a great game." (Max Delbrück, third from the right, at a conference at the Niels Bohr Institute, Copenhagen, 1934. Courtesy Victor Weisskopf and AIP Niels Bohr Library)

The Student and the Stars

THE DELBRÜCK FAMILY

All Delbrücks are related to one another, and all are numbered. Max's number, 2.517, indicated that he was the seventh child of the eldest child of the fifth child of Gottlieb Delbrück, who lived in Halle from 1777 to 1842. The numbering starts with Gottlieb (Delbrück #2) and his elder brother Johann Friedrich, the number-one Delbrück.

In Max's words, "the name 'Delbrück' was quite well known; too well, in fact, for the comfort of a youngster, who felt ambiguous about being recognized and identified immediately as a member of this clan, but not for himself" (A). A few examples will illustrate the intellectual world into which Max was born. Johann's son Rudolf, carrying the number 1.1, played a decisive role in forming the North German Confederation. This union turned out to be an important stepping-stone in setting up the German Reich. When this was achieved in 1871, Rudolf was appointed to head the chancellor's office in Berlin. Rudolf von Delbrück possessed all the positive features of a typical Prussian bureaucrat of the old school. A firm Lutheran traditionalist, he was driven by Hegelian ethics and devoted his person to "the generality incorporated in the State." At the same time, he was surprisingly modern, an advocate of free trade.

Other Delbrücks had made their mark in other areas. Heinrich, a cousin of Max's father, became president of the Supreme Court. Another cousin, Berthold, is considered one of the founders of Indo-Germanic language studies. Max's father's brother, also called Max, became a well-known industrial chemist; he founded and directed a large research institute in Berlin supported by the fermentation industry.

Max's father, Hans Delbrück, of North German Protestant stock, was a professor of history at the University of Berlin; he also edited a monthly journal called *Preussische Jahrbücher (Prussian Year-books)* that published essays in the humanities and political science. This magazine ran a political column of sixteen printed pages, which Hans Delbrück wrote for many decades. He addressed these aggressive articles to the liberal intelligentsia. A collection of the essays was published in three volumes in 1918, entitled *Krieg und Politik (War and Politics)*.

Hans Delbrück was a prolific historian. His pre-1920 lectures were published as a *Weltgeschichte (World History)*. As a historian, he was mainly concerned with the history of the art of war in its political ramifications, completing seven volumes on that topic. In the introduction to the fourth volume, Hans Delbrück explained the purpose of his collection: "This work has been written not for the sake of the art of war, but for the sake of world history. If military men read it and are stimulated by it, I am pleased and regard this as an honor; but it was written for friends of history. . . ."

Hans Delbrück's specialty was what he called *Sachkritik:* criticism of sources, in which he found corrupted descriptions of battles. For a wide range of historical periods, he would research the logistics available and examine the terrain, to check extant descriptions. However, his main interest lay in general ideas and tendencies rather than in the minutiae which had crowded the pages of earlier military histories (Craig, 1943).

Hans Delbrück was also politically active. For five years, he lived as a tutor at the court of Crown Prince Frederick. After that, Max's father was elected for one term to the Prussian Parliament (1882–85) and for one term to the Reichstag (1884–90). He wanted to find out how the parliamentary system worked; he never made a speech, although he was always very critical of imperial decisions. Indeed, a book exists with the title *Hans Delbrück: A Critic of the Wilhelminian Era* (Thimme, 1955).

Max's mother Lina came from South Germany (Fig. 1). Her father was Carl Thiersch, the surgeon general of the allied armies in the Franco-Prussian War in 1870 and a professor of surgery at Leipzig, where she grew up. Her grandfather was Justus von Liebig (1803–1873), the world-famous chemist. Appointed Professor of Chemistry at the age of twenty-one in Giessen, Von Liebig helped

1. *Max's parents, Hans and Lina, in Berlin. Courtesy of Emmi Bonhoeffer.*

make Giessen a center of chemistry study. He introduced mineral fertilization and developed theoretical and agricultural chemistry. Justus von Liebig is considered one of the founders of scientific chemistry; his analytical results contributed to establishing that carbon is the element common to all organic compounds.

EARLY YOUTH IN BERLIN

Hans and Lina Delbrück had seven children, Max being the youngest. There were four girls and three boys. Before Max was born on September 4, 1906, the Delbrücks decided that a family that large needed a house, which they built in the Grunewald in Berlin. This affluent suburban area of Berlin seemed to be crawling with big faculty families. Max wrote of his home and its neighborhood,

> My parents had the house built according to their design, simple but spacious, and had moved there at the time of my birth. The house, while within the residential area, was only a block away from the city railroad and a freightyard. A convenient location, since the city railroad and

streetcars were the main conveyances at that time; on summer nights when the windows were open we could hear time dropping away with the city railroad puffing and starting, and the freightcars clanging back and forth. (A)

The move to the Grunewald had been initiated by Lina Delbrück, who was described by Max as possessing "a very sweet disposition." He remembered her as "completely loyal" to his father, "willing to take directions." The Delbrück house, though, was mainly built through her influence. Due to Lina, also, the house in later years became a gathering place for friends and relatives. What especially stood out in Max's mind were the huge table where large crowds would take meals, and the beautiful garden. Max always remembered his early life in Grunewald as a paradise—a paradise that was lost in August 1914 with the outbreak of World War I.

Max was eight years old at the time. He recalled the start of war:

The dam broke and swept away four decades of peace. The vaunted equilibrium of power collapsed like a house of cards. The fury of war, at first greeted with hysterical optimism and self-righteousness, soon changed Europe into a nightmare of death and destruction. Although in retrospect the transition marked by August 1914 was not as radical as that, to me it has always appeared as the most absolute break that ever occurred in history.

Vaguely I remember soldiers marching through the streets in these August days and we children rushing there and handing them red currants we had picked in our garden as they were marching along. (A)

FATHER AND SON

Young Max resented being recognized and identified immediately as a member of the Delbrück clan rather than for his individual merit. Even as a small child, he looked forward to eventual renown. At a family gathering when Max was six, someone remarked that famous men never have famous children. Such a statement did not sit well with Max. His sister Emmi recalls that he assumed his characteristic grin and replied, "Just wait!"

Max also felt ambiguous about his famous father, who was approaching sixty when Max was born. Hans Delbrück was well known by then; nevertheless, he continued to maintain an exhausting schedule, working daily from 8:00 a.m. until midnight, with a solid siesta after the noon meal (Max took such siestas in his later years). During his adolescence, Max developed an uneasy relationship with his father, manifesting subconscious hatred and jealousy mixed with admiration and respect.

This complex attitude toward his father lingered long past Max's childhood. As an adult, Max was once asked at a party, "Why do you work so hard?" Before Max could respond, a friend of his answered, "He works for the woman he loves." In retrospect, Max found this statement truer than the friend had intended. To Max, the response did not mean that he worked for his wife, but for his mother—to outshine his father.

THE NEIGHBORS IN BERLIN

Max's neighbors in Grunewald consisted mainly of distinguished academic families. The Delbrücks were especially close with two such families. One of these was the family of a psychiatry professor, Karl Bonhoeffer. His eldest son, Karl Friedrich, became Max's first mentor and life-long friend. The second Bonhoeffer son, Klaus, married Max's youngest sister, Emmi. The youngest of the three Bonhoeffer brothers, Dietrich, became a theologian. It is mostly through Dietrich's opposition to the Nazis that the name Bonhoeffer rose to fame in Germany. For openly resisting Church support of the Nazis, Dietrich was put in a concentration camp and murdered in 1945. His brother Klaus suffered the same fate.

The second family close to the Delbrücks was that of Adolf von Harnack. A professor of theology, Harnack was Lina Delbrück's brother-in-law. He had first held faculty positions in Giessen and Marburg. When he was offered a job at the University of Berlin, the Supreme Council of the Evangelical Church announced its opposition, only to be overruled by Emperor William II. In 1890, Harnack was elected to the Prussian Academy of Sciences in Berlin. At the dawn of the new century, his lectures entitled *What Is Christianity? (Das Wesen des Christentums)* created much excitement. This book is now considered a classic historical study in the field of theology; it is regarded as the one text that "more directly than any other represents so-called liberal Protestant theology" (Pauk, 1968).

In 1906, Harnack accepted an appointment as director of the Prussian State Library, the largest and most important library in Germany. He later became widely known as the founder of the

Kaiser Wilhelm Gesellschaft. Named its first president in 1911, he continued at this post to the end of his life in 1930.

In contrast to the universities, which had to give their primary attention to teaching, the Kaiser Wilhelm Gesellschaft, or Society, was organized to launch scientific institutes devoted to research. The Gesellschaft rapidly established several research facilities, chiefly in the natural and medical sciences. Almost immediately, the institutes won worldwide recognition and influence. After World War II, the organization developed into today's Max Planck Gesellschaft.

As a young man Max never liked Harnack, and as a scientist he never approved of the Max Planck Gesellschaft, for he believed that research should remain in the universities. During the 1970s, when he was on the board of this organization, Max even inquired jokingly into the possibility of dissolving the whole Max Planck Gesellschaft.

It is an amusing coincidence that Max Planck, the great physicist, lived down the street from the Delbrücks and the Harnacks when Max was a child. In fact, Max used to steal cherries from the garden of the discoverer of quantum theory. The Delbrücks and the Harnacks knew nothing of the scientific advance that began in their neighborhood.

Within the Harnack, Delbrück, and Bonhoeffer families, there were twenty-two children; there was also continuous intellectual ferment. On Sunday nights, the families met regularly. Often the adults discussed current affairs while young Max and the other children listened. The fathers never discussed scientific matters, or religious questions. They commented on needed political and social reforms and bemoaned the dangers of excessive nationalism. These running commentaries continued throughout World War I and during the postwar period of revolution, inflation, and impoverishment. The only difference was that, after 1914, the Sunday-night meetings no longer included dinner.

THE IMPACT OF WAR

The war impinged upon Max's peaceful childhood existence. According to his later recollections, "My memories start with the war years: hunger, cold, substitute teachers, social pressure to . . . engage in dismal patriotic war games. . . ." (A). The reference here is to a Pan-German youth movement which it was proclaimed a national duty to join. Though the organization of this group was similar to that of Baden-Powell's Boy Scouts in England, the spirit of the group differed totally. Enormous social pressure was exerted to achieve wide-scale participation. These German Boy Scouts met every Sunday morning from eight to one o'clock. In sun, rain, or snow, they marched, sang songs, and engaged in fairly rough war games. Max dared not admit for more than a year that he hated the sessions; on the contrary, he staunchly pretended enthusiasm. Even sixty years later, he remembered this as one of the great lies of his youth. His mother eventually managed to get him out of the movement without too much disgrace.

One historian has described the mood in Europe at the outbreak of the war as a "curious compound of uncomplicated patriotism, romantic joy in the chance of participating in a great adventure, and naive expectation that, somehow or other, the conflict would solve all of the problems that had piled up during the years of peace" (Craig, 1981). On the German side, an active Pan-German League exerted pressure for annexations, of Belgium, for example, and for the creation of satellites such as Poland. The leadership of the Pan-German League worked hard to mobilize the German intelligentsia, particularly the university professoriate. Hans Delbrück undertook the task of organizing a countermovement. Stressing the defensive character of the war, he favored a simple demonstration of German might. He pointed out in a petition in July 1915 that "the greatest prize that victory can bring will be the proudly achieved assurance that Germany need not fear even a whole world of enemies and the unprecedented testimony to our strength that our people will have given to the other peoples of the earth" (Craig, 1981).

These views were supported by Max Planck, Albert Einstein, and Adolf von Harnack. But the petition against German expansionism attracted many fewer signatures than a memorandum from the Pan-German League calling for annexations. This defeat of the Delbrück petition helped to confirm the German people in their irrational military expectations. Their fantasy world collapsed in October 1918.

Max remembered his mother's desk laden with "many photographs of all the young men of our friends and relatives who were killed in action," including his oldest brother, Waldemar, who died

in 1917. Armistice Day coincided with his father's seventieth birthday. Hans Delbrück, a patriot to the core, felt this to be "the darkest
day of his life. During the remaining ten years of his life he spent
much of his time and effort on two parliamentary commissions,
one of them investigating the origin of the war and the other one
the reasons for the collapse" (A). For Max,

> the principal change brought about by the end of the war was the return
> of good teachers from the war, especially my teachers of Greek, Latin,
> German and mathematics. I was then twelve, so this transition coincided
> with the transition to high school. In Germany at that time this meant
> also transition to long pants and formal address. You were expected
> from now on to be a serious young man and not a rowdy. Fine! We all
> did respond. Some of us, like me, on the timid side, responded with
> relief. (A)

The teachers were excellent, but at the same time quite disillusioned. The enthusiasm and patriotism that greeted the war at first,
and the great trust in Prussian virtues and the emperor's leadership, had all given way to a deep mistrust of the state and its
representatives. After all, the emperor William II had simply fled
across the border into Holland.

MAX AS A YOUNG MAN

The German government that followed the abdication of the last
Hohenzollern is known as the Weimar Republic. Weimar Germany
has often been described as a cradle of modernity. Those who
created Weimar culture believed they were living in a new age in
which everything had to be developed afresh. The Republic gave
freedom and encouragement to its artists and intellectuals, resulting in the poetry of Benn, the plays of Brecht, and the novels of
Hesse and the brothers Mann. This period also saw the rise of
expressionism in art and literature, Bauhaus architecture, the physics of relativity, psychoanalysis, and atonal music.

At the end of the war, revolt against authority was commonplace.
Though German youth had not won a new sense of assurance or
direction, they did realize that they had to define their own values.
They could trust neither the Church nor the state. Max was typical
in turning from politics and the Church, becoming apolitical, de-
tached from society. To the end of his life, Max consistently avoided

politics—whether national, international, or academic. Max was also typical in setting an independent path. He was a teenager when the Weimar Republic began. At this age, he "tried to find out who [he] was, i.e., [he] was groping for an identity as we say since Erikson's famous book *Young Man Luther*" (A).

Max gained such identity by turning to a field his relatives, friends, or neighbors knew nothing about. He had been very quick with calculations in school: in astronomy and in mathematics, he soon found an independent self. His romantic hero was Johannes Kepler, whose picture hung over his bed. At the age of twelve, on the evening of his confirmation, Max swore to himself that he would remain faithful to the stars.

At this point, he acquired his first mentor: Karl Friedrich Bonhoeffer (1899–1957) was seven years older than Max, the oldest of the seven Bonhoeffer children who lived a block away. Karl Friedrich was the only scientist Max knew then; he would later become an outstanding physical chemist, being named director of the Max Planck Institute for Physical Chemistry in Tübingen after World War II. When Karl Friedrich investigated Max's interest in astronomy, he found it childish, but he also found Max willing to learn. The older Bonhoeffer boy taught Max what Max later referred to as the difference between vague talk and real insight. Max felt Karl Friedrich taught him an important lesson: When you do something, do it well or not at all.

This friendship had consequences beyond refining Max's efforts. Bonhoeffer also led Max to value the knowledge-seeking goals of science. In Max's words,

> this guidance, together with the other educational influences and the Protestant moral principles that, largely unspoken, were strongly encouraged in the family, consolidated at that time in me the feeling "The Truth shall set you free." This motto, which happens also to be the motto of the California Institute of Technology, is strangely ambiguous. As it happens, it is taken from the fourth gospel, the gospel according to St. John, and occurs there in the context of the heated discussions between Jesus and the Pharisees. The truth to which it refers is the claim of Jesus to be the Son of God, a very far cry from scientific truth, and yet, at depth, related. The meaning in which I wish to use the motto here refers to the romantic feeling that the pursuit of scientific truth, of poetic truth or of mystic truth, is ultimately far more important and influential in shaping man's fate than the power game of those with political aspirations who try to change the world directly. . . . (A)

A LECTURE ON ASTRONOMY

When Max finished high school in 1924, his Greek teacher chose him as valedictorian. Max focused on Kepler as his topic. While preparing for the talk, he found in his father's large library a collection of speeches by the former director of the Berlin Astronomical Observatory, a man by the name of Archenhold. This included a speech on Kepler from which Max lifted a few passages without acknowledgment. Fifty years later, Max remained embarrassed about this, especially when he remembered how well his speech was received.

The teacher suggested that Max, in preparing the speech, examine some of Kepler's books in their original editions; these were available in the rare book collection at the university library. The university then permitted youngsters to use such books in the reading room. Years later, Max recalled that "to see and handle these old books, 300 years old, was a tremendous experience, especially when I found in one of them some of the speculations about the celestial harmonies expressed in terms of musical notes. It showed that Kepler literally thought in terms of a 'heavenly dingdong' at that time, rather than in terms of abstract mathematics" (A). Kepler thus followed ancient Pythagorean thought.

This suggestion by his teacher instilled in Max an approach to scholarship: he experienced what it meant to go back to original sources instead of relying on current interpretations in the secondary literature. Very early Max learned to establish an original line of inquiry.

A STUDENT OF ASTRONOMY

The boy's astronomical interests in the 1920s were tolerated indulgently by his parents. His observations had to be made at night, and the only place in the house to set up a telescope was a little balcony adjacent to his parents' bedroom. Max's alarm was set for 2:00 a.m.; unfortunately, he was the last one to hear it. The family patiently put up with this disruption. Max retained his interest in the stars through high school, and determined to study astronomy at the university.

All through high school, German pupils were subjected to drills and to rigid structure. Once they were past these hurdles, the freedom at university was infinite: there were no term papers, no finals, and no grades. Max thoroughly welcomed, and enjoyed, the German university system.

He began his university career as an astronomy student in Tübingen in the summer of 1924, when he was not yet eighteen. His first professor was Hans Rosenberg, who was then introducing astrophysics, a new science, in Germany. An astrophysicist measures not the position of stars but their spectra, using sophisticated physical devices. Thus, one would have to learn both mathematics and physics to master astrophysics. Max did so.

As a student, Max developed many habits that would stay with him. First, he gained much through speaking with older students, and became committed to the value of "talking to people a good deal" (A). He also formed a characteristic style for listening to lectures:

> At that time I developed the habit of not taking any notes at lectures but watching the professor and trying to follow the mathematical argument with some understanding. When the professor made a little mistake, with a plus or a minus sign or a factor 2, I did not point that out directly but waited ten minutes until he got entangled and then pointed out, to his great relief, how he could disentangle himself—a great game. (A)

The 1920s were not the best years for a young man to be studying astronomy in Germany. In Max's later view, German astronomy had been ruined by the overambition of the astronomers who lived in the second half of the nineteenth century and were following up on the tremendous triumph of Friedrich Wilhelm Bessel. Bessel had succeeded in measuring the first parallax of a star in 1838. Determining the parallax of a star means detecting the difference in direction of the star as measured from two points on the earth's orbit around the sun. If one views an object from two different points that are not on a straight line with it, then from each vantage point the object seems to be in a different place compared to a given background. This is called parallax. For instance, if one looks at the hand of one's watch under different angles, one's reading of the time may vary by several minutes. For a successful parallax measurement, one had to measure values of angles of a few tenths of an arc second. This technology was not available until the first part of the nineteenth century.

When Bessel achieved his breakthrough, it was important not only from a technical viewpoint but also from a scientific one. An elementary implication of the Copernican hypothesis is that if the earth moves around the sun, one should be able to see the stars wiggle back and forth. This immediate and obvious inference was not verified for three hundred years—between 1543, when Copernicus published his revolutionary work, and 1838, when Bessel took his measurements.

After Bessel, German astronomers took great pride in refining position measurements, not only by using parallaxes but also through determining the proper motions of the stars; they soon made elaborate position catalogues of the stars. The first one, called *Bonner Durchmusterung (Bonn Survey)*, was completed in the 1850s. In the 1880s, astronomers started a second collection of positions, believing that repetition of these measurements fifty years later would provide an accumulation of a vast number of proper motions from which astronomers could derive the structure and dynamics of the galactic system. By the time Max was a student, the fifty years were nearly over. But by then, astronomers discovered they had overestimated the accuracy of the 1880 measurements. Thus they would have to wait another fifty years before they could really possess sufficiently accurate motions: they would have to discard all the earlier measurements and start afresh with the improved methods. This focus, Max felt, destroyed German astronomy. Young astronomers sat up every night for hours measuring the transit of stars, which in addition to being of little value was "a painful thing to do because you couldn't even heat the building, otherwise the air was not quiet enough" (A). The effect on the intellectual quality of German astronomy was devastating.

About the time Max became a student of astronomy, several scientists decided it was time to apply more sophisticated physics to the discipline. Hans Rosenberg in Tübingen was one, as was Hans Kienle in Göttingen. Göttingen was the more exciting place of the two because of David Hilbert's presence there among a galaxy of other mathematicians. Göttingen ranked first in physics, thanks to Max Born and James Franck and the unfolding of quantum physics.

THE LESSON OF ASTRONOMY

Max's choice of astronomy as a profession turned out to be a very bad one. Germany had been foremost in the field during the nineteenth century when astronomers were making catalogues with precision instruments to measure star positions, proper motions, and parallaxes. But this approach had now run into a dead end. The new astrophysics was developed by American observatories, the leader among them being the Mount Wilson Observatory in Pasadena. The theoretical approaches were pioneered largely in England. What was developed in Germany was not astronomy, but quantum physics.

Max had heard about the breakthrough in physics in 1925 while still in Berlin. He went to Göttingen in 1926 as an astronomer, but his allegiance changed when physics became sensational and his own thesis on theoretical astronomy did not work out. Max had wanted to develop a theory explaining novae—new stars which suddenly increase their luminosity tremendously. Such an event had occurred in the summer of 1926, and a number of astronomers had followed the spectral changes of the star in moderate detail. Max remembered that one astronomer got so excited about what he saw that he sent a telegram to the German monthly astronomical bulletin stating NOVA PROBLEM SOLVED. STAR EXPANDS AND BLOWS UP.

Max set himself the task of developing a theory to explain what may have happened inside such a star—how the temperature, pressure, and ionization vary from the periphery to the inside. He soon came to realize that in order to conduct such stellar dynamics, he would have to read and understand the theoretical papers on stellar interiors and atmospheres that had been published by Arthur S. Eddington and E. A. Milne. But these were written in English; since Max had not yet learned English, such a task was impossible in any reasonable time frame. Even his astronomy professor could not read them, although he was keenly aware that stellar dynamics led to the astronomy of the future. Max gave up on astrophysics and moved to quantum mechanics.

Although Max did not complete his thesis, he did feel that this aborted experience in astronomy contributed much to his scientific development. Many years later he described this gain in a 1970 lecture on Copernicus. It was his second astronomical lecture. In order to prepare it, Max went back—as he had done fifty years before—to the original sources, visiting the rare book collection at the UCLA Medical School to understand more about Copernicus.

In his lecture, Max pointed out that when Nikolaus Copernicus proposed the sun as the center of the planetary system nearly five hundred years ago, he thereby initiated a revolutionary scientific paradigm. Copernicus favored this theory for its simplicity as well as for its explanatory power. At first, however, the application of heliocentricity did not lead to a resolution of complexities; it must have been exceedingly distressing for Copernicus to attempt to decipher such an amazingly complex celestial clockwork. The ultimate clue, the solution to the confusion, was still missing. Kepler finally resolved the dilemma early in the seventeenth century by positing the elliptical movement of the planets. Copernicus' reliance on simplicity as a criterion for scientific explanation proved fruitful, as has happened often in recent centuries.

Max in his later years liked to summarize seminars, books, or more complicated interactions into a "take-home lesson." The take-home lesson from his study of astronomy was that one's trust in simplicity to explain the mechanisms of nature will be rewarded.

MAX IN WEIMAR POLITICS

Max studied astronomy and physics between 1924 and 1930. In these years following World War I, Germany was a republic in a state of crisis. Each of the fourteen years of the Weimar Republic was troubled. The political shock of the military defeat in 1918 was followed by the humiliation of the peace terms in Versailles in 1919. When to these was added the harrowing experience of enormous inflation, the Germans who had accepted the transition from Empire to Republic lost their faith in democratic processes.

In his autobiographical notes, Max described three incidents that exemplified for him the political climate of this first German Republic. The first example involved an assassination. In 1922 Walter Rathenau, then Minister of Foreign Affairs, was killed one morning on his way to his office, a block away from the high school that Max attended. Rathenau, a Jew, was murdered by members of a secret, nationalist, anti-Semitic group. The school, which Max attended for twelve years, still stands and has recently been renamed the Walter Rathenau School.

Max's own words convey his second memory from this period:

The first President of the Republic, Ebert, a Social Democrat, had died, the new president was to be elected. The Conservatives and Nationalists

put up Hindenburg, a national legend because he had happened to be the
top general at the beginning of the war when things were going well on
the eastern front. Everybody by 1925 who had any knowledge of the true
history of this phase of the war was aware of the fact that Hindenburg
had made no contribution. My father thought that the choice of Hindenburg was a terrible one. (A)

Liberals feared that Hindenburg's election could provide a grave
blow to German democracy.

Max remembered a small part he played in this election cam-
paign. Though the description is not in complete agreement with
the facts, his story is nevertheless of interest:

The two candidates opposing Hindenburg, who happened to be called
Marx and Luther, were both acceptable to the more liberal people. During those years Kurt Hahn had befriended my father. This friendship
had started during the last year of the war when Kurt Hahn had become
the private secretary and speech writer for Prince Max of Baden. This
prince had become Reichskanzler during the last month of the war,
entrusted with the unenviable task of liquidating it. Hahn was a highly
interesting character who like Leo Szilard much later believed and in
fact proved that a single individual person without office, without party,
without any academic degrees even, can move history. The scion of a
wealthy heavy industry family, the only Jewish one in the Ruhr district,
I am told, he struck out on his own. He must have been about forty years
younger than my father and often came to visit him, quite often at eleven
in the evening, unannounced, and then read to him the pages of the
prince's memoirs which he was ghost writing. In 1925 Kurt Hahn persuaded my father that the only chance to block off Hindenburg would
be if either Marx or Luther stood back in favor of the other, and that my
father should write a personal appeal in the form of a letter to both of
them and that these letters should be delivered personally to them at
their headquarters with me as the messenger. My father agreed. The
letter was drafted, typed by one of my sisters, and Hahn and I and his
driver set out in his car to do the delivery. Both candidates were in
conference at the time I arrived, so I could only make my personal
delivery indirectly. Of course, neither of them did step back. And accordingly during the summer, when I was a student in Bonn, one morning
walking to the university, I read the shocking headline: HINDENBURG
ELECTED REICHSPRESIDENT. (A)

The facts of this incident vary slightly from Max's recollections.
Hans Luther was then the chancellor; he was not running for president. Max's father, estimating that Luther might have a better

chance than Wilhelm Marx to beat Paul von Hindenburg, asked Luther to resign as chancellor and Marx to withdraw from the race. Neither complied. In the election, Marx lost to Hindenburg by fewer votes than a third candidate, a Communist, received. The liberals had proven ineffectual and German democracy stood threatened with Hindenburg at the helm.

The third incident occurred during the winter. Max was a student at the University of Berlin, and,

> in addition, had taken an [unpaid] job as a research assistant at the tower telescope of the Einstein Foundation, located on the premises of the Potsdam Observatory. This tower telescope had been conceived by Dr. Erwin Freundlich, who was an enthusiast about Einstein and about general theory of relativity and [had] organized several expeditions to observe the deflection of starlight during total eclipses of the sun. He had also [raised] the money to construct this tower telescope, the purpose of which was to confirm the red shift of the spectral lines in the sun's spectrum, another prediction of [Einstein's] general relativity [theory]. On the first day of this job, Freundlich . . . decided that it would be polite to introduce me to the Director of the Observatory, . . . [who] at that time was Prof. Hans Ludendorff [a brother of General Erich Ludendorff]. . . . (A)

When Max was introduced to Hans Ludendorff, an astronomer of distinction who had also published papers on Mayan astronomy, Ludendorff withdrew his hand immediately upon hearing the name Delbrück. He asked whether Max was the son of the historian. When Max said he was, Ludendorff turned on his heel, slamming the door of his office. He later accused Erwin Freundlich of a deliberate slap in the face in hiring the son of the man who had insulted his brother.

Hans Delbrück had indeed reviewed General Ludendorff's memoirs in an extremely critical fashion. This review, which Max never read, had been published as a pamphlet and sold widely at newspaper stands (H. Delbrück, 1920). It took several weeks for Max to resolve the impasse with the general's brother.

For Max, these three incidents "illustrate[d] the political passions that seethed through all the years of the Weimar Republic, and led to its collapse in 1933" (A). Since Berlin was the center of such political storms, and since Max was uncomfortable in such a political climate, he loved to escape whenever possible to the country. He found a retreat in Stawedder, a hamlet near the Baltic sea, north of Lübeck.

FAMILY LIFE

Max later wrote of his experience in Stawedder:

> To the decisive influences of my early adolescence belongs Stawedder. There an older cousin of mine, Paul Carrière, a composer and violinist, and his wife, Lotta, a pianist, had bought a cottage and had remodeled it very attractively and cozily. The war and the inflation of 1923 had wiped out their savings so they had a hard time making ends meet. Paul, a lovely and gentle soul, was quite useless at earning money. Lotta went around in the neighboring villages and gave piano lessons to the farmer's children. In addition they took in boarders, including my sister Emmi and me, on many of our school vacations. To get to Stawedder from Berlin was quite a challenge, taking several changes of train and finally a long march on foot. Several times I carried on this long march my 2-inch telescope weighing 20 pounds. Of course we traveled 4th class on the slow trains. Once it turned out that at one of the stations where I had to change, the connection was a fast train, not having 4th class, and cost some extra fare which I didn't have. The country station was quite deserted between trains, except for the station-master and maybe one other person. They became intrigued by the boy who carried a heavy telescope so I set it up and together we looked at the landscape, upside down, of course, since it was an astronomical telescope. Eventually our intimacy became sufficient for them to advance me the few Marks to cover the extra fare.
>
> We developed a great passion for life with the Carrières in this idyllic country environment and the cottage with a cow and a pig and a grandmother and three children, a wonderful grand piano, leisurely chess games with Paul under the flowering pear tree, or evenings in the living room and a wide variety of music. This attachment to the Carrières has persisted through the decades and now extends to the third generation, not less intense than at its inception. (A)

This contentment in Stawedder illustrates a characteristic of Max's that was to persist throughout his life. He loved being surrounded by family. In later life, the concept of family would include close scientific colleagues as well as relatives.

Religion carried little force for him. He could never come to grips with the theological writings of his brother-in-law, Dietrich Bonhoeffer. In the Delbrück family, it was assumed that one believed in God and that one was Protestant. Although Max developed a deep fondness for religious music, such as Bach's St. Matthew Passion, he never felt any religious inclinations. He considered religion to be a fantasy woven around a nucleus that contained some truth; the combination functioned to supply the strength needed in life. For Max, science fulfilled a similar function. The difference was that in religion one depended on God, while science left one on one's own. Max preferred to remain on his own. A poem by Goethe that Max was familiar with illustrates the distinction well:

> *Wer Wissenschaft und Kunst besitzt,*
> *Hat auch Religion;*
> *Wer jene beiden nicht besitzt,*
> *Der habe Religion.*
>
> (Whoever possesses Science and Art
> Has also Religion;
> Whoever does not possess these two,
> He should have Religion.)

Education in Germany in the early part of the twentieth century was not judged according to one's knowledge of the Bible, but rather by one's knowledge of Goethe's works. Members of the Bonhoeffer family, especially, knew large parts of Faust I by heart. It was customary then for every commencement speech in Berlin to cite Goethe's writings. Max's own family instilled very early in him an appreciation for the great German poet.

At the end of World War II, Max made a substantial effort to acquire the collected works of Goethe from his father's library. He then proceeded to house the more than twenty volumes in the bathroom of his house in Pasadena. This unusual and provocative placement did not indicate contempt for the books. They have been carefully restored, the faint titles rewritten in Max's own hand, and he loved to read in them.

He clearly also enjoyed the conventional reactions of visitors who encountered the Goethe set in the bathroom. Max always took pleasure in provoking others, and awaiting their response.

Theoretical Physics—
The Intellectual's Game

THE EXCITEMENT IN PHYSICS

The revolution in physics at the beginning of the twentieth century was sparked by two men, both of whom lived in Max's neighborhood while he was a student. In 1900 Max Planck, a professor at the University of Berlin, and in 1905 Albert Einstein, then a technical expert at the Swiss Patent Office in Bern, obtained results that indicated there was something wrong with the foundations of classical physics. At first, the discovery of quantum physics in 1900 and of special relativity in 1905 did not have a strong impact. But by the mid-1920s, the dawn of a new era was obvious to all scientists.

While Max was working in the Einstein Tower in Berlin in the winter of 1925–26, he heard rumors that a breakthrough had occurred in the understanding of atomic phenomena, in the form of a quantum theory. Werner Heisenberg had conceived the new theory in the spring of 1925, while he was recovering from hay fever in Helgoland. It had subsequently been worked out mathematically by Heisenberg, Max Born, and Pascual Jordan in Göttingen in the same year (Heisenberg, 1969). In the spring of 1926, Werner Heisenberg, then twenty-five years old, was invited to present a seminar at the University of Berlin. Its Physics Department was the premier center of physics in Germany—with Max Planck, Albert Einstein, Max von Laue, and Walther Nernst all on faculty. It was here that Planck had discovered the quantum of action; here, too, Einstein had formulated the general theory of relativity in 1916. The central event of department life was the general Physics Colloquium—a tradition that could be traced back to the nineteenth century, when it was first organized by the great physiologist Hermann von Helmholtz. The entire Physics Faculty would attend the Colloquia.

Max heard about Heisenberg's scheduled seminar, and planned to attend. Going into the building just before the lecture, he saw Einstein and Nernst together at the entrance door. Nernst was very excited; he mentioned that he had read Heisenberg's reprint and asked Einstein: "Do you think there's anything to this?" Max heard Einstein's reply: "Ja, ja, I think it's a very good paper, very important." Max realized that something fundamentally new was being presented, and he felt excited to be at the source. The seminar room was packed, with a hierarchical seating pattern. In the front row sat Nobel Prize winners: Einstein, Planck, and Von Laue. In the second row were the associate professors, and so on down, with standing room only for the others. Max later admitted that he did not understand Heisenberg's seminar, but he was nevertheless impressed.

What was new in this presentation of atomic phenomena was that Heisenberg for the first time turned from describing the orbits of electrons in the atom, since—as he pointed out—they cannot be observed. He had developed a theory that included only observable quantities like frequencies and amplitudes, using them to represent the electrons. Though Max was only nineteen years old, and though he did not apprehend the physical and mathematical details, he perceived an important aspect of this revolutionary physical theory: that it relied on simplicity and beauty as criteria for truth (Heisenberg, 1969). In the coming years, the new theory succeeded in accounting for atomic phenomena. Max thus encountered here another instance in which reliance on simplicity led to scientific advances.

MAX IN GÖTTINGEN

In the mid-twenties, the major excitement in science took place in physics; one of the centers for this activity was Göttingen. This was mainly due to Max Born (1882–1970), and his two young assistants, Wolfgang Pauli (1900–1958) and Werner Heisenberg (1901–1976). Max Delbrück arrived in Göttingen in the summer of 1926 to study astronomy. During the three years he stayed, he found himself more and more attracted by the quantum picture of the world. When his thesis project in theoretical astrophysics fell through in

1928, Max turned to the new quantum mechanics, breaking with astronomy. Fortunately, as he recalled, the language of the new physics "was definitely German" (A).

This shift was given additional momentum through a long-term friendship that began in Göttingen in 1928. Victor Weisskopf, a graduate student from Vienna, was two years Max's junior. Both were eager to acquaint themselves with the new quantum theory, so new it was not regularly offered in classes yet. They organized private classes to catch up with the new developments.

Both young men had emerged from families steeped in the humanities. Both entered physics with zero knowledge of quantum mechanics, and both liked to talk. They spent their evenings learning the new theories, an activity they called "Eine kleine Nacht-physik" (A Little Night Physics).

Later, Max and Weisskopf would get their Ph.D.s under Max Born. Born and Niels Bohr (1885–1962) were the senior quantum physicists, with Heisenberg, Pauli, Jordan, and Paul Dirac forming the younger generation. One of Born's assistants in Göttingen was Paul Ehrenfest, who later taught a course on the new quantum theory that Max and Weisskopf attended. Ehrenfest liked to encourage his students to see that "the most stupid questions might be the most interesting ones." His favorite sentence was "Physics is simple, but subtle."

Victor Weisskopf remembers Max in Göttingen in the late twenties as a shy person, who liked to present himself as an experienced man of knowledge. But Max was always willing to learn. Both shared a sense of humor they put to work in creating their own "comic physics." They spent an exciting time in Göttingen in the atmosphere created around Born while the revolutionary quantum physics was being developed. Max always remembered vividly the mysterious atmosphere of Göttingen's forests at night as he and his colleagues looked at the stars, wondering if they could really grasp what Einstein had presented to the physics world: a universe that had no limits but was still finite. Max questioned his friends as to the inconceivability of such a universe.

Aside from Victor Weisskopf, Max teamed up with Maria Mayer and Edward Teller. At that time, Robert Oppenheimer and Norbert Wiener were also in Göttingen. Max recalled the student body as a collection of oddballs. He noted the mannerisms Oppenheimer had adopted so that he would fit in:

I found out at an early age that science is a haven for the timid, the freaks, the misfits. That is more true perhaps for the past than now. If you were a student in Göttingen in the 1920s and went to the seminar "Structure of Matter" which was under the joint auspices of David Hilbert and Max Born, you could well imagine that you were in a madhouse as you walked in. Every one of the persons there was obviously some kind of a severe case. The least you could do was put on some kind of stutter. Robert Oppenheimer as a graduate student found it expedient to develop a very elegant kind of stutter, the "njum-njum-njum" technique. Thus, if you were an oddball you felt at home. (D 81)

The years 1928 and 1929 in Göttingen opened the doors for Max to enter not only the quantum world but also the world of poetry and philosophy. He was introduced to these fields by the person he liked to call his "second most important mentor" (he considered Niels Bohr the most important one). In Göttingen, Max developed a very close friendship with Werner Brock, who was six years his senior. Officially a philosophy student, Brock had come to Göttingen to write a monograph on Nietzsche. He had studied with Karl Jaspers in Heidelberg, later becoming an assistant to Martin Heidegger in Freiburg. However, Brock was much more than a student of philosophy. He had also completed medical studies, with special attention to psychiatry. In addition, he had a profound knowledge of art history and of many other subjects—except science. With strong views on so many areas, he stimulated Max's reading.

Brock encouraged Max to study the poetry of Rainer Maria Rilke, especially the *Duino Elegies,* and Shakespeare's sonnets. Max was shocked at first by the *Duino Elegies,* but he continued to think about them over the next fifty years. In 1980, he began preparations for a lecture on the eighth of these elegies, sadly a lecture he was too ill to complete.

Although science was the only field in which Werner Brock was not an expert, he encouraged Max to continue, even though at the time Max did not feel that his physics thesis constituted exciting science. He sensed that the important ideas in theoretical physics had been introduced by others and only details remained to be filled in. Max felt at the time that he was not doing well, but hoped that something would come along for him.

Werner Brock's influence on Max, as significant as it was, lasted only a limited time. Brock had long been mentally unstable, with episodes that resembled schizophrenia. Since he was half Jewish, he had to emigrate to England in 1933, a move that turned out disastrously for him. He was unable to adapt to the English language or the English approach to thinking and writing. He attempted to lecture on Jaspers and Heidegger in English and to convert the lectures into books, but his German style lacked appeal.

Brock became quite paranoid. He returned to Freiburg after the war, never to recover. When he was institutionalized, Max visited him several times. Brock recognized Max and held quiet conversations with him. Aware of his condition, he knew that he would never leave the mental institution.

THE NEW PHYSICS

The story of atomic physics begins in the 1890s. By then, X-rays had been discovered; as a consequence, a breakthrough was achieved to a new level of understanding of the behavior of matter. Classical physics had led previously to a general description of material properties. One could characterize the properties of matter in terms of material constants, such as specific heats, specific conductivities, refractive indices, and numerous others. With such constants that themselves were not susceptible to further interpretation, one could predict the behavior of matter. Before the 1890s, only a vague notion of atoms and molecules existed; physicists were not convinced they would ever materialize.

This situation changed radically in the last decade of the nineteenth century. With the new discoveries, atoms proved real and of defined size, weight, and structure. They contained electrons as subunits, and their positive charges were shown to be subject to change in radioactive decay. It was now possible to begin to interpret the periodic system of elements. There seemed to be no reason why, within a few years, scientists should not know exactly how atoms are structured, how electrons move about, or how the positive charge is distributed.

There were a few dark clouds: It was not clear how electrons could exist. If they were point charges, their field energy would be infinite. If they were finite in size, what kept them finite? Also something seemed wrong both with classical mechanics and with the electromagnetic theory of radiation, each of which proved inapplicable in the interaction between radiation and matter.

When analyzing the radiation of a black body, Max Planck discovered at the turn of the century that the energy of radiation comes only in discrete packages called quanta. With v being the frequency of radiation, a quantum would have the energy $E = hv$, with h being Planck's constant, also known as the quantum of action. Einstein followed this up five years later, using the photoelectric effect to convince physicists that there were indeed such quanta in light. These photons, as they were called, caused an annoying problem: Light was known to spread like waves; how could it then consist of microscopic particles, the photons?

Each of these arguments implied that classical mechanics and electrodynamics could not both hold in the interaction between radiation and matter. Einstein immediately appreciated the gravity of the challenge, commenting in his autobiographical notes (Schilpp, 1949, p. 45): "It was as if the ground had been pulled out from under one, with no firm foundation to be seen anywhere."

The crisis of physics was brought into clear focus in 1911. The resolution owed debts to both theory and experiment. Planck had introduced his quantum hypothesis to do away with what is called the ultraviolet catastrophe. According to classical physics, the spectrum of radiation emitted by a black body would increase without limit at high frequencies, and the total energy inside a box in thermal equilibrium would be infinite. These predictions were clearly nonsense. Planck found that by introducing his famous quantum hypothesis, he could obtain the correct result for black-body radiation. In 1911, Paul Ehrenfest showed that the quantum hypothesis formed the *only* way to avoid the ultraviolet catastrophe.

In the same year, it was discovered that each atom had a nucleus. In Ernest Rutherford's laboratory, a beam of positively charged particles was directed at matter in order to analyze its scattering. Most of the beam particles were not deflected at all, but some were scattered very strongly. This result allowed Rutherford to propose an atomic model that resembled a solar system in miniature. A heavy nucleus is surrounded by electrons that move around in their orbits. The nucleus would be small compared to the size of the orbit, but large compared to the electrons. If we visualize the nucleus as the sun, the electron's orbit would resemble Neptune's path rather than the Earth's.

Rutherford's suggestion implied that in a hydrogen atom, a single electron would circulate around a proton. But an electron is a charged particle; in classical physics, such a moving charge would radiate energy. It would lose speed as a result of the energy it radiates, and be caught by the proton. So classical physics could not account for the most obvious property of an atom: its stability.

Thus either Rutherford's model of the atom or the laws of classical physics had to be wrong. In this situation, Niels Bohr took a bold step. He saw that the model was obviously correct since the experiments proved it. As a consequence, one had to change the laws of classical physics. In the years following Rutherford's discovery, Bohr expanded the bounds of physics to incorporate the new observations.

In 1913, Niels Bohr displaced the paradox of the atomic stability with a schizophrenic theory. He calculated the orbits of the electrons in the hydrogen atom classically, as one would calculate the orbit of a satellite. He then picked out a few such paths, declaring the electron to be stable in each such quantum state. Only if the electron jumps from one state to another would it radiate energy. This would involve a discontinuous process. Bohr's calculations for the size of the released quanta coincided with experimental findings, constituting the first triumph of quantum theory.

As it turned out, it took more than a decade to achieve the comprehensive formalism called quantum mechanics. This was accomplished by Werner Heisenberg in 1925 and by Erwin Schrödinger in 1926. It was not until 1927 that a deeper understanding arose of the necessity of this formulation. This understanding, called the complementarity argument, provided the motivation for Max's subsequent scientific work.

QUANTUM PHENOMENA

Quantum theory, as it was worked out in 1926, constitutes an absolute break with classical concepts. It describes a world with phenomena that are radically different from normal experience in the everyday world. In order to illustrate the new features of reality that arise when one is dealing with events on the microscopic scale, Niels Bohr conceived of the two-slit experiment. Max described this "thought experiment" to introduce quantum behavior in the last course he taught at Caltech in the late seventies, drawing heavily on the presentation given by Richard Feynman in his physics lectures (Feynman, 1963).

Thought (*Gedanken*) experiments were originally introduced to physics by Einstein. In a thought experiment, one considers a pertinent conceptual configuration designed to make theory easier to think about. One applies a theory to experimental conditions without actually doing experiments, in order to see the implications of theory in concrete terms.

Bohr's argument unfolds in four steps. Each experiment uses a standard set-up involving two slits (or holes) in a wall and a source of waves or particles. From the source, particles or waves are emitted to the wall; they move through the holes, are sometimes deflected by striking the side of the hole, and arrive at a backstop (or absorber) where a detector is installed. In the first experiment, a gun fires bullets which the detector counts. If only one slit is open, an intensity distribution illustrated schematically in Figure 2 by either P_1 or P_2 will be obtained, depending on which slit is open. If both slits are open, these two distributions are superimposed without interference: $P = P_1 + P_2$. Such is the behavior of classical particles.

In the second experiment, the set-up is immersed in water (Fig. 3); water waves are generated through the slits and the detector determines their intensity. Since waves show interference, the intensity determined when both holes are open is not the sum of each individual intensity. Instead, an interference pattern is recorded.

In the next experiments, electrons are fired through the two holes

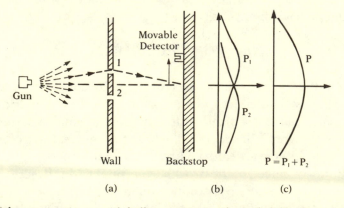

2. Particle gun experiment with bullets, measuring the probability that a bullet will pass through a hole and arrive at the backstop. Here there is no interference; the probability with both holes open equals the sum of the probabilities with individual holes open. From Feynman, Leighton, and Sands, The Feynman Lectures on Physics, vol. 1. Cambridge, Mass.: Addison-Wesley Publishing Co., 1963. Reprinted with permission.

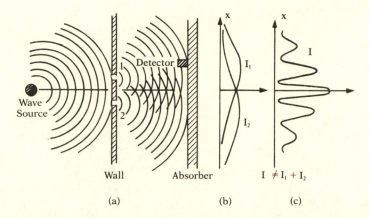

3. *A wave experiment. The detector at the backdrop measures the intensity of the wave motion proportional to the square of the height of the wave motion. The waves interfere with each other; the individual intensities do not add up to the intensity observed with both holes open at once. From Feynman, Leighton, and Sands, The Feynman Lectures on Physics, vol. 1. Cambridge, Mass.: Addison-Wesley Publishing Co., 1963. Reprinted with permission.*

in the wall. Here the detector is a device that registers electrons; it is hooked up to a loudspeaker, and each incoming electron is recognized by a click. Thus, at the backstop the electrons are registered as particles (Fig. 4): they will be diffracted (like waves) but nevertheless distributed similarly to the bullets if only one hole at a time

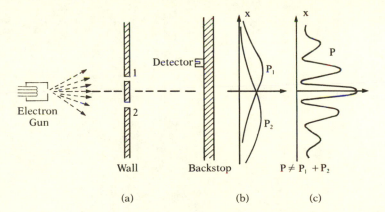

4. *Electron gun interference experiment. The sum of the individual probabilities that an electron arrives at the backstop when each hole is open separately does not equal the probability with both holes open at once. Electrons thus behave like waves, showing interference. From Feynman, Leighton, and Sands, The Feynman Lectures on Physics, vol. 1. Cambridge, Mass.: Addison-Wesley Publishing Co., 1963. Reprinted with permission.*

is open. However, if both holes are open, interference is observed. The electrons thus demonstrate their wave nature. Electrons are quantum objects.

With a little thought, one can see that this proves not only that matter (in this case the electrons) has wavelike properties, but more specifically that the way an object appears to a subject (the observer) depends on what the subject is doing. For instance, the observer listening to the clicks at the backstop will find electrons recorded there as particles. But the situation is different at the wall where no one can detect the electrons directly; at the wall it is not known through which slit an individual particle passes on its way to the backstop. In fact, one must conclude that the electrons move as waves through the holes in the wall, since they show interference at the backstop (Fig. 4(a)). In other words, it does not make sense to ask if an electron went through hole 1 or hole 2.

This situation changes if the observer does look to decide through which slit a particular electron went. In principle, one could employ a light source near the wall in such a way as to determine the path of the electron as it passes through one slit or the other (Fig. 5). However, this cannot be done without influencing the behavior of the electrons in such a way that no interference will be observed. (Fig. 5(a)). The subject would have forced the object to behave like a particle at the slit.

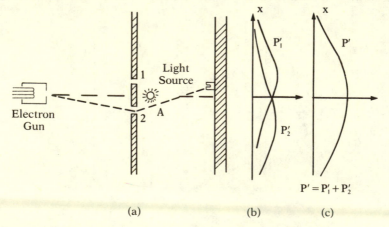

5. *Electron gun experiment with a light source between the holes, to allow the observer to see which slit the electron goes through. The interference disappears now; the unobserved electrons behave differently from the observed electrons. From Feynman, Leighton, and Sands,* The Feynman Lectures on Physics, vol. 1. *Cambridge, Mass.: Addison-Wesley Publishing Co., 1963. Reprinted with permission.*

THE COMPLEMENTARITY ARGUMENT

In 1927, Niels Bohr invented the term "complementarity" to be able to discuss such paradoxical situations confronting physicists. In complementarity, he wanted to reconcile the various contradictory points of view that seemed to be emerging in physics. Waves and particles are two classical concepts which apply to atomic phenomena only in a mutually exclusive way. Such sets of concepts are called complementary if together they provide a complete description of the phenomena under consideration and if the use of one of them excludes the simultaneous application of the other.

In the description of atomic events, Bohr noticed that we are dealing with individual processes which are inseparable from their respective experimental arrangements. Various such arrangements are mutually exclusive of each other, leading to observations on the same phenomenon that are complementary to each other. He introduced the term "complementarity" in September 1927 at an international physics congress in Como, held to commemorate Alessandro Volta, who had died a hundred years earlier. Bohr had conceived of the term earlier that year while skiing in Norway (Moore, 1974).

This feature of atomic physics, expressed in Bohr's complementarity argument or in the more popular "uncertainty relations" by Heisenberg, proved shocking to the physics world. It was in fact so much of a shock that Albert Einstein never got over it. For the rest of his life, Einstein attempted to return physics to the classical picture of just one physical reality. He postulated that if one cannot get at the full reality with present methods, then presumably other methods must exist to do the job (Shilpp, 1949).

On the other hand, Bohr insisted that this limitation to the classical picture of reality was not a preliminary stage to be replaced by a return to classical notions, but served as an advance over them. He believed that physics had arrived at a new dialectical method to cope with a totally unexpected feature of reality. Bohr continued to elaborate and restate this argument in innumerable lectures until his death in 1962. In the years following his initial presentation of the complementarity argument, he tried to generalize the notion, to attribute to it a universal meaning, and he wanted to formulate the consequences of this insight for other sciences.

MAX IN PHYSICS

The principal features of the new atomic theory and the fundamental equations of quantum mechanics had all been worked out by investigating simple systems like the hydrogen atom and the harmonic oscillator. Niels Bohr often reminded coworkers that the laws of quantum mechanics had been found by guessing, which was possible only because physicists had available a simple atom in hydrogen to test their ideas against reality. A simple system was needed for such success at the outset; later, one could understand more complex systems.

The nucleus of a hydrogen atom was a proton, an elementary particle with a positive charge. The more complicated helium and lithium were supposed to have two or three protons in their nuclei. The question then arose as to the forces that keep these protons in position, since protons should repel each other through their positive charges. The structure of the nucleus provided a fundamental challenge for physicists in the late twenties. Another promising field was the study of complex systems with many interacting particles. It was not clear whether the equations of quantum mechanics would allow an understanding of such systems. It is true that a more and more complicated mathematical machinery had to be invented to handle complicated molecules, but could one—by starting with the fundamental Schrödinger equation—derive the properties of molecules from first principles? Would the fundamental equations hold in complex cases without need for refinement or addition?

In 1928, Eugene Wigner had introduced the concepts of group theory, especially of the so-called permutation group, to classify the spectral terms of multi-electron systems (Wigner, 1927). This new development fascinated Max. He had not seen any application in theoretical physics of group theory, especially the theory of finite groups; he thought he could easily catch up and work on this without competition. Since the group concept provides an adequate mathematical language to define the symmetry of the system under investigation, Max expected that by exploiting the symmetry of a molecule to study its multi-electron system, he could ease the mathematics involved considerably.

By the time Max became interested in group theory, Richard Courant and David Hilbert had just published a famous treatment of the methods of mathematical physics. This quickly became "the bible of theoretical physicists" (A), yet it nowhere mentioned group theory. It happened, however, that a slim book on group theory by A. Speiser appeared in the same series of treatises published by Springer. Although it lacked any reference to physics (Speiser, 1937), Max found that

> by reading the elementary parts of Speiser's book, you could understand Wigner's paper and get ahead of everybody else in the game. I did so and noticed that Wigner in his paper had omitted one or two proofs that he should have given. I met Wigner at the swimming pool and asked him about it. He immediately said, "You are absolutely right. Why don't you publish these proofs?" I said, "Why don't you?" He said, "No, I can't. It's too small a matter but I will help you write this little paper." And so he did. It was a marvelous experience in the course of which I learned the actual handling of the wave-mechanics of multi-electron systems, the elements of group theory and how to write a paper. Looking now at this paper, entitled "Ergänzung zur Gruppentheorie der Terme" [D1], I find that I have a hard time understanding it since it contains a fair amount of algebra involving representations of the permutation group by matrices and the reducibility of these representations in special cases.
>
> Intrinsically, this paper contains not a single new idea. However, it had momentous consequences for me personally. It drew the attention to the young graduate student of astronomy who had jumped into this new field, especially the attention of W. Hielter who had just come to Göttingen to be an assistant of Max Born, and indirectly the attention of Max Born. As a result I got a little job as teaching assistant of Born. This was my first paid job which boosted my self-confidence enormously even though the pay was minimal, as were my duties. (A)

This job eventually launched Max on a thesis project in theoretical physics, suggested by Walter Heitler. The project was to explain in semiquantitative terms why the covalent bond between two lithium atoms was so much weaker than the homologous bond formed by two hydrogen atoms. The latter had been analyzed very successfully by Heitler and Fritz London the year before (Heitler and London, 1927). Their quantum mechanical theory of the hydrogen molecule explained, reasonably satisfactorily, the strong bonding of the two atoms in terms of what was called an exchange integral.

Quantum mechanics had abandoned the classical picture of electron orbits envisioned by Bohr; instead, the best one could do was to calculate the probability for detecting an electron in a given region around the nucleus (Fig. 6). All locations with nonzero probability constituted the orbital of an electron. In the Bohr atomic world there were various such orbitals, depending on the charge of the corresponding nucleus. Hydrogen possessed only the simplest such orbital, which was called 1s; it had one electron moving around. In more complicated atoms, electrons would team up in even numbers in what were known as shells. The shell closest to the nucleus was named the K shell; it could hold two electrons.

The Bohr atomic model could thus explain the periodic system of the elements, in which atoms were ordered according to their mass. They would be listed with an order number, starting with the number 1 in hydrogen and moving all the way up to 92 in uranium. It was first thought that this number would indicate how many particles the corresponding nucleus contains. Instead, it turned out to yield the number of such particles with charge.

Although atoms could thus be accounted for by the new quantum mechanics, molecules would prove more difficult to explain. In the case of hydrogen, one would expect certain effects to result from efforts to bring two such atoms into closer contact; that is, the negatively charged electronic orbitals would cause a repulsion and prevent the formation of a molecule. But Heitler and London could show theoretically that in the case of hydrogen, the two 1s orbitals would fuse and a new orbital would form, keeping the two nuclei (protons) at a distance in a stable position. This union would constitute the hydrogen (H_2) molecule (Fig. 7).

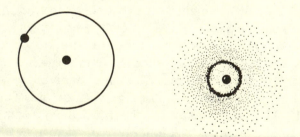

6. *Two ways of representing a hydrogen atom. On the left, the electron's orbital is depicted as a definite line. On the right, the orbital is defined as the speckled region representing the relative probabilities of where the electron is likely to be found. Adapted from Biological Science, 4th edn., by William T. Keeton and James L. Gould. New York: W. W. Norton & Co., 1986.*

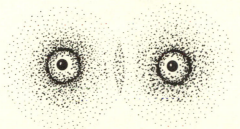

7. Covalent bonding of a hydrogen molecule. Chemically, the proximity of electron clouds should lead to repulsion. Quantum mechanics explains that the overlapping electron clouds form a strong new configuration, characterized by sharing of electrons. Adapted from Biological Science, 4th edn., by William T. Keeton and James L. Gould. New York: W. W. Norton & Co., 1986.

The Heitler-London bondage is a feature of quantum mechanics, not an invention to explain the chemical bond. Yet the theory corresponded exactly with all known chemical facts, suggesting to physicists that they could account for many types of aggregates from first principles. Some physicists, excited by this success, predicted chemistry would soon become merely a branch of physics. In her biography of Niels Bohr, R. Moore quotes a young physicist from Copenhagen who believed that in a matter of years, the new theory would explain the nucleus and thus would complete physics. From then on, chemistry and biology would be accounted for. This, of course, has not been the case.

Max slowly climbed the ladder of complexity, concentrating on lithium for his Ph.D. thesis. Lithium is the next higher homologue of hydrogen. Its innermost electrons form the filled K shell. Beyond that, it contains one electron in a so-called 2s orbital instead of a 1s orbital as in hydrogen. The question of concern was why the two lithium atoms, which have electrons in the 2s orbital, do not form a strong bond in the same way hydrogen atoms do.

Max realized that this project, suggested by Heitler, "involved no essentially new idea" (A). He felt, though, that it would give him experience in handling wave mechanics and might yield "a quick thesis. It turned into somewhat of a nightmare since it involved horrible six-dimensional integrals in polar coordinates around two centers. Since this was before the day of computers, the question turned around finding analytical expressions for these integrals, or suitable approximations to such" (A). Eventually, in the summer of 1929, Max did produce a thesis on the lithium molecule; he characterized it as "acceptable though rather dull" (A).

This retrospective judgment rests on the fact that Max had not produced what he really wanted to, namely, "a new idea." Simply carrying out calculations was not his ambition; he was interested in developing theory. Such a mandate was difficult in physics during 1928 and 1929. The most important concepts in quantum theory had already been formulated before Max entered the field. Some hope was left, nevertheless: another storm of ideas might become necessary to explain the atomic nucleus. Bohr and Heisenberg believed—one might even say they hoped—that by moving even deeper into the heart of matter, from the atom to the nucleus, one would generate another revolutionary theory.

THE FIRST TIME ABROAD

During the summer of 1929, Professor John E. Lennard-Jones from Bristol University had come to Göttingen to learn German and to master the new quantum mechanics. He was especially interested in chemical bonding. At the end of the summer, Lennard-Jones had learned neither German, since all had attempted to improve their English with him, nor the new quantum mechanics. Since he had a postdoctoral fellowship to offer, he asked for someone to join him in Bristol. To Max's surprise, Max Born recommended him and Lennard-Jones hired him to come to Bristol in the fall of 1929, even before Max could complete his final oral examinations. This offer provided a marvelous opportunity for Max to go abroad and to learn English (Fig. 8).

Max liked to point out that before his move to Bristol, he decided to concentrate on learning English phonetically rather than working on grammar or syntax. This strategy paid off wonderfully, since it meant that from the beginning whatever he learned of the English language remained relatively free of the heavy German accent which has plagued the vast majority of Germans abroad. He was proud that his English pronunciation was nearly perfect; he never tired of correcting his Spanish, Japanese, or German students, postdocs, and friends.

8. Passport photo of Max in 1931, before leaving for England. Courtesy of Manny Delbrück.

Max adapted to the English-speaking world with characteristic rigor. For his first three months in Bristol, he opted to live with a parson's family and share all his meals with them. He wanted to spend most of his time improving his English. During this period, he gave a few lectures on quantum mechanics to the local physics group. He later commented on this stage of his acclimatization:

> There was nobody in Bristol who spoke German, so the language training was rather intense and I enhanced it deliberately by not buying an English-German dictionary but only the Concise Oxford English Dictionary. Thus looking up any word generally led to looking up five different words. It was a great moment when this process of looking up words ceased to diverge into an infinite search. (A)

Bristol was an attractive university in the sense that the Physics Department there had just received a large sum of money for expansion; it had hired several young experimental physicists, mostly from Cambridge. Bristol offered good facilities, and the group was in high spirits. One member was C. F. Powell, who rose to fame as the discoverer of the pi meson, receiving the Nobel Prize in 1950. Powell and Max, who later shared a room in Bristol, became very close friends (Fig. 9).

In December 1929, Max had to return to Göttingen to take his final oral examination. To his utter consternation and to the surprise of many of his friends, Max failed. He attributed this failure to both "foolish naivete and arrogance" (A). His friends had told him that at the final oral examination the faculty asked only trivial and elementary questions, and that there was no need to study. Max took them at their word. He went into the examination

blithely without having looked at a book. We had four examiners: in my case the astronomer Kienle, the mathematician Courant, a theoretical physicist, Born, and an experimental physicist, Pohl. Pohl did not think kindly of theoretical physicists and I in fact had zero knowledge in his field but I thought I could afford to flunk one of the four subjects. Therefore as he examined me I was unable to answer any of his questions but remained very calm and simply said "I don't know." (A)

9. Max and his roommate, C. F. Powell, in Bristol (1932). Courtesy of Victor Weisskopf.

Pohl became more and more irate. Max flunked, and had to repeat the exam the next year. When he returned to Bristol, his colleagues were amused and puzzled that the young man recommended to them as so highly talented should have met with such indignity.

Immediately after this disaster, Max Born feared that his former student might commit suicide. The Borns invited Max home that evening to calm him down. Max remembered the invitation, but not any severe depression.

During the next three months in Bristol, Max's unhappiness grew. This was due neither to the examination nor to his uninspired relations with Lennard-Jones; it resulted instead from his realization "that the English culture was really totally different from the German one. There seemed to be almost no overlap between the books read, the thoughts . . . the subjects discussed, the activities undertaken, the style of personal relations, [or] the view of history" (A). From his German point of view, his "English friends appeared to be unbelievably naive and unintellectual." Yet in another three months, a complete reversal took place. Max suddenly came to appreciate the English way of life, becoming quite an Anglophile. He felt reborn: "I began to appreciate the radical difference between the two cultures and ever since then I have felt that the English experience, language and culture-wise, has been the greatest enrichment of my life" (A).

THE FIRST ROCKEFELLER FELLOWSHIP

After a year in Bristol, Max felt the need to work in an environment which had more to offer in theoretical physics. With recommendations from Max Born and Karl Friedrich Bonhoeffer, he obtained a one-year fellowship from the Rockefeller Foundation. Max decided to spend the first half in Copenhagen at Niels Bohr's institute and the other half in Zürich with Wolfgang Pauli.

He arrived in Copenhagen in February 1931, to be taken in hand immediately by George Gamow. Gamow, two years Max's senior, was from Leningrad. Since only a few Russian scientists had come to Germany following the 1917 Revolution, Gamow was a rather exotic figure. Max came to Copenhagen with barely any notion of what he would work on. Gamow instructed him as to the status of theoretical nuclear physics. The Russian physicist had been very successful in the quantum mechanical interpretation of the radioactive α-decay. He had worked out two refinements (Gamow, 1931), namely, the interpretation of the so-called long-range α-particles and the fine structure of the energy of some atoms that showed α-decay. The goal was to understand the structure of an atomic nucleus. Gamow had shown in 1928 that one could, in a first approximation, compare such a nucleus to a droplet of water. He observed that the forces binding the particles in the nucleus would be very short in range, such that they acted only between two neighboring nuclear constituents.

Scientists had considered the radioactive decay observed in some elements as the appropriate window for examining the nucleus. The Curies had discovered at the turn of the century that during such a process, particles are emitted from the nucleus; two of these particles were identified by Ernest Rutherford: the α- and β-particles. In some cases, two types of α-particles appeared that differed in their energy and thus in their range. This produced the fine structure of the spectrum. Short-range particles were always more prevalent than long-range particles.

Gamow's theory of the nucleus postulated that it could be in one of two states: low energy (ground state) and higher energy (excited states). The transition between them would be accompanied by emission of radiation (γ-rays). He then interpreted the long-range particles as decays not from the ground state, but from an excited state of the nucleus emitting the α-particle. Gamow also tried to understand the fine structure of the α-decay by comparing the probabilities for the different states. Since some experimental data existed relating the transition probabilities of α-decay and γ-radiation, the question arose as to whether the data sufficed to draw inferences about the nature of the forces that held the nuclear constituents together.

At that time, the summer of 1931, the neutron had not been discovered and electrons were supposed to float around in the nucleus in some nebulous manner. Nevertheless, Max and Gamow together made "an heroic attempt" to arrive at estimates for these transition probabilities, or at least at upper and lower limits (D 4). This search taught Max a great deal about the status of experimental and theoretical nuclear physics; furthermore, it enabled him to conceive the first speculative idea that he considered original.

PHYSICS BEFORE 1932

The year 1932 proved a key one for Max and for physics. These facts are not unrelated: 1932 marked the discoveries of the deuteron, the neutron, the positron, and the disintegration of nuclei by artificially accelerated protons. Before the discovery of the neutron, there was no way to construct a nuclear model without electrons. Such hypothetical "nuclear electrons" were studied and described by Gamow, who in his "Constitution of Atomic Nuclei and Radioactivity" (1931) used special signs (intended as skull and crossbones originally) to set off speculative passages, indicated to the reader where the content was not firmly grounded.

The phenomenon that provided evidence in favor of electrons involved the β-decay. In this process, electrons appear to emerge directly from the atomic nucleus. Even more puzzling was the observation that these emitted electrons possessed differing energies, since this would imply "that the internal energy of a given nucleus can take any value within a certain continuous range" (Gamow, 1931). Since all the observed nuclei before or after the β-emission seemed identical, Gamow concluded:

> As was pointed out by N. Bohr, we must reckon with the possibility that the continuous distribution of energy among the nuclei is fundamentally not observable, or, in other words, has no meaning in the description of the physical processes. . . . *This would mean that the idea of energy and its conversion fails in dealing with processes involving the emission or capture of nuclear electrons.* (Gamow, 1931, author's italics)

By the end of 1930, Wolfgang Pauli had begun to suggest the possible existence of a neutrino which would eventually resolve the difficulties. But in 1931, Bohr and Heisenberg were willing to abandon the hard-won conservation laws considered to form the pillars of science. They seemed to welcome a failure of quantum mechanics at nuclear dimensions, hoping to arrive at an even more fundamental theory that would require radically new conceptions of space, time, and causality. Such ideas and hopes were debated at the 1931 International Conference on Nuclear Physics in Rome, which Max attended.

Then, in January 1932, the neutron was discovered, and Heisenberg conceived of the idea that the nucleus consisted of protons and neutrons. The electrons seen in the β-decay were now considered to have been created in the process of emission. To explain the stability of the atom had required new fundamental laws; no model in the existing body of physics sufficed. However, to explain the stability of the nucleus, physicists simply needed to apply quantum theory in the appropriate way.

This discovery proved greatly disappointing to those attracted to physics for its revolutionary character, its openness to new ideas. If quantum mechanics was the definitive theory of atoms and its parts, it meant that physics might now be reduced to a straightforward application of fundamental equations to particular problems.

SUMMERTIME IN COPENHAGEN

Gamow was not the only influence on Max's career in this period. Niels Bohr had a far greater effect. The Danish physicist presented a striking contrast to Max Born, Max's teacher in Göttingen. Max recalled that whereas Born knew all the answers, it was Bohr who would ask the right questions. Max Born had been an excellent advisor, yet he always kept considerable distance between himself and his students. Max remembered only a cool relationship. On the other hand, Niels Bohr was like a father; being at his institute felt like being in a family—a family, of course, with an international character.

There were eleven postdoctoral fellows when Max arrived, each capable of conceiving important new ideas and doing the appropriate calculations quickly and reliably. A rule existed at the institute that in giving a seminar, one could use any language except one's mother tongue. Max could have used English, but he tried to learn Danish, becoming fluent enough to write letters in Danish and to read Søren Kierkegaard in the original.

The center of the activities was the Institute for Theoretical Physics on Blegdamsvej in Copenhagen. Built in 1920, it attracted brilliant theoreticians from the start. Here Werner Heisenberg and

Niels Bohr worked out what is now called the Copenhagen interpretation of quantum theory, in 1926 and 1927. (The institute is
now known as the Niels Bohr Institute.)

While working in Ernest Rutherford's laboratory in Manchester,
Bohr had published "On the Constitution of Atoms and Molecules"
in 1913. The ideas in this paper opened the path for an understanding of the structure of matter. Bohr was awarded the Nobel Prize
in 1922, one year after Albert Einstein.

By the time Max came to Copenhagen in 1931, the principles of
quantum mechanics had been worked out. Bohr was deeply concerned now with the implications of this new theory, which seemed
to involve epistemological problems. He was attempting to formulate the lesson the new physics could offer to other sciences such
as biology. It has to be assumed that Max's interest in the life
sciences originated from his time at the Niels Bohr Institute in the
memorable summer of 1931.

To Niels Bohr, quantum theory gave the final word on atoms. His
main interest was to understand the epistemological consequences.
After all, the world of quantum mechanics functions differently
from the world one lives in. It is an "Alice-in-Wonderland world"
(D 111). Atomic particles no longer follow trajectories, but vanish
down rabbit holes to escape deterministic descriptions. They cannot even be marked for later identification. Quantum systems behave totally distinctly from objects of normal size. To deepen this
quantum epistemology and to follow this conceptual line of research seemed to Max more attractive than to study the fine details
of the theory, as physicists at Max Born's institute in Göttingen
were doing.

Shortly before the end of Max's first stay in Copenhagen, Victor
Weisskopf came to work in Bohr's institute. Max picked him up at
the station, announcing that the best thing about Copenhagen was
its beautiful women and that he had already invited some of these
over for that evening. After some reluctance Weisskopf joined the
party—fortunately so, since he met his future wife, Ellen, there.

ANOTHER FAUST

Until 1937, Max returned to Copenhagen at every possible opportunity, especially to participate in the spring conferences that were

attended by many former fellows (Fig. 10). The 1932 meeting ended in a party "with a skit that has gone down in history" (A). That year marked the one hundredth anniversary of Goethe's death, and Max set out to write a parody of *Faust* with the help of Carl Friedrich von Weizsäcker (Weizsäcker, 1983).

In Goethe's play, God agrees to a bet with the devil, Mephistopheles. The bet concerns Faust, who has studied and worked deeply in all sciences and still is disillusioned that he does not understand how things work. If Mephistopheles can satisfy Faust's search for understanding, then the devil can take him; otherwise Faust will go to heaven. The parody was written in such a way that God was meant as Bohr, Faust was transformed into Ehrenfest, and the devil was represented by Pauli. Max was having fun with the difference in personality between the two good friends Pauli and Bohr. He made up a disagreement about physics to center the action around. In the end, God and the devil agree to a compromise.

The whole play was performed by the junior physicists; Leon Rosenfeld portrayed Bohr, and Felix Bloch acted as Pauli. Max appeared between the scenes to explain the original *Faust.* His only active part came when he announced the "Klassische Walpurgisnacht" in which a number of witches meet. In Goethe's *Faust,* excitement fills the stage here. In the parody, nothing happens, causing Faust (alias Ehrenfest) to complain. Max went on to explain that in a "classical" Walpurgisnacht, there was no interaction with

10. Niels Bohr, Max Born, and Max in Copenhagen (1934). Photo in possession of Manny Delbrück. Courtesy of the Niels Bohr Institute of Copenhagen.

the observer (the audience). He proposed going ahead with a "quantum theoretical" Walpurgisnacht; all agreed.

The disagreement between God and Mephistopheles involved the future of physics. The devil is ready to throw out concepts such as mass and charge, while God objects that this would leave nothing at all. At the end of the play, a particle is presented that seems to solve the problems Bohr and Pauli were concerned with. This is the neutron, which has no charge but maintains mass nevertheless.

THE NEUTRON

This particle, the neutron, had just been discovered in 1932 by James Chadwick, changing the whole concept of nuclear structure. Now the nucleus was known to contain two particles. These possessed the same mass, one having a positive charge (the proton), the other being neutral (the neutron). Max theorized that the particles of unit mass could also exist in states of much higher charge, positive or negative. These states of higher charge, if they were realized inside an α-particle, would explain the very high binding energy of the nuclear constituents. Now we know that this binding energy is caused by a different force, called the strong force.

Max showed a draft of this speculation to Bohr and Rutherford during a visit to Copenhagen; and they in turn recommended its publication in *Nature* (D 5). This speculation excited quite some interest for a short while, until more elementary particles—the mesons—were hypothesized by Yukawa to provide a better explanation of the strong forces (Yukawa, 1982). In the late 1930s, all considered Yukawa's theory as correct. Today, scientists believe one has to penetrate to more elementary constituents of matter, in fact to quarks, to provide a truly correct explanation of the nuclear forces.

It is significant that headway on the problem of the nuclear force was made not by Western physicists, but by Yukawa of Japan. He came up with the proper model of the strong interaction shaping the nucleus by assuming the existence of a new elementary particle. Among the European scientists, "the whole climate of opinion at that time was against new particles" (Dirac, 1978). Even Wolfgang Pauli waited three years to publish his idea of a neutrino.

By the time Max's speculation was published, he was back in Bristol after having spent a scientifically unproductive, but personally rewarding winter with Wolfgang Pauli in Zürich in 1931–32. Pauli did not treat Max with his well-known, merciless rudeness; he made an exception, and the two became life-long friends. It is possible, though, that Max's lack of success in Zürich and a developing feeling of mathematical inferiority accelerated his move to biology.

George Gamow once reported in an interview that Pauli scolded Max, at a cocktail lounge in Zürich that winter, about some of his physical ideas. Possibly Max began to realize that physics was too complicated for him (von Meyenn, 1982). Max reported never having been able to fill a page with mathematical formulas without making several mistakes. He simply felt that he could never match colleagues like Victor Weisskopf and Wolfgang Pauli. Max favored ideas more than calculations; he would welcome a field whose fundamental problems would allow nonmathematical solutions.

THE RETURN TO BERLIN

Max's second stay in Bristol, in the summer and fall of 1932, ended three years abroad that enormously widened his horizons. He then had to decide where to settle. Max later recalled having had the choice of becoming an assistant to Wolfgang Pauli in Zürich, or of moving back to Berlin to join Lise Meitner, who was investigating the structure of matter by physical means at the Kaiser Wilhelm Institute for Chemistry. He opted for returning to Berlin, explaining his reasons in a letter to Bohr dated June 1932:

> I have accepted Lise Meitner's offer to go to Dahlem as her "family-theoreticist" in October largely because of the neighbourhood of the very fine Kaiser-Wilhelm-Institute für Biologie, to which I am entertaining friendly relations. Also here in Bristol, I have continued my biological learning chiefly in the domain of Botany, and I have even gone to the length to take part in a botanical excursion to the South coast, with the Bristol Botany professor and some students. This was very instructive, as the professor happened to be a passionate teacher.

This is, indeed, the earliest time one can document Max's interest in biology. There is no hint that he thought about biological problems or that he educated himself in that field while he was in Göttingen or during his first stay in Bristol. Max's biological learning seems to have begun during the six months he spent in Copenhagen, most likely induced by Bohr's complementarity argument or—to be more precise—by Bohr's hope that this idea might have general epistemological significance. Bohr's presentation of this view to the public in the summer of 1932 coincided with the start of Max's activity in biology.

During the fall of 1932, Max moved back to Berlin, to spend the next five years in the house he had grown up in (Fig. 11). From there, he could commute quite easily to Dahlem—a residential suburb similar to Grunewald but several miles farther out—where the institutes were located. Between 1932 and 1937, Max made a gradual transition from theoretical physics to biology.

Officially, of course, he worked in physics; but he set aside as much time as he could to develop Bohr's conjecture and to gain a footing in biology. His physics activities and his biological activities of the next five years—Max's final years in Europe—are best understood if described separately. The next chapter treats the physics side.

11. *Max and his mother, Lina, in Berlin (1933). Courtesy of Emmi Bonhoeffer.*

The Berlin Years

DELBRÜCK SCATTERING

In physics, Max was confronted with rather complicated problems. In his Berlin years, he published two important papers, each quite different from the other. The first consists of a half-page addendum to a paper by Lise Meitner and Heinz Köster (D 6); the second, written with Gerd Molière, is a very long and learned application of the new quantum mechanical concepts to statistical physics. The short addendum contained no calculations; it brought Max's name into physics textbooks for the introduction of what is now known as "Delbrück scattering." In the long paper, the authors conducted complex calculations, only to reach a disappointing conclusion. They could not clear up the problems of statistical mechanics they had intended to, and hardly anyone noticed the treatment. In later years, quite a few physicists tried their hands on the same problem, coming up with similar answers (Fig. 12).

Max conceived of the scattering mechanism that is named after him when he tried to interpret the results of experiments that had been undertaken in Lise Meitner's laboratory. In 1932, Meitner was investigating the scattering of "hard gamma rays" (D 6). "Hard" rays consist of high-energy radiation, energy usually measured in units of electron volts (eV). One eV is the kinetic energy an electron gains if one volt is applied. Hard γ-rays have energies that are millions of times that high; in the experiment under consideration, Thorium C was used as a source: the γ-rays leave this element with 2.6 MeV.

When directed toward lead, the γ-rays emitted by Thorium C (high-energy protons) are scattered by the electrons that move around in their orbitals. The collision between these outgoing pho-

12. *Max with several of his Berlin colleagues at a Physics Colloquium in Copenhagen (1936). Seated in the front row are (left to right) Niels Bohr, Paul Dirac, Werner Heisenberg, Paul Ehrenfest, Max, and Lise Meitner. Courtesy of the Niels Bohr Institute of Copenhagen.*

tons and the electrons produces what is called Compton scattering. This effect, discovered in 1922, made it convincingly clear that electromagnetic radiation also has particle-like properties.

Meitner and Koster, however, found additional scattered components whose intensity was much too high to be interpreted as coherent scattering due to the bound, inner-shell electrons. It seemed as if the γ-rays had actually been scattered by the nucleus itself. To be more precise, since the nucleus contains positively charged particles, it is surrounded by a strong electromagnetic field. The light would then be scattered by interacting with this nuclear field. This phenomenon is called Delbrück scattering even though Max never presented any calculations. In the addendum to Meitner and Koster's paper, he pointed out how the observed scattering component may have been produced. The mechanism he suggested differs somewhat from the interaction implied by the modern usage of the term "Delbrück scattering."

In 1933, Max proposed that the observed additional scattering which now bears his name could be explained by a new theory Paul Dirac had recently proposed (Dirac, 1928; Dirac, 1957). Dirac had reformulated the Schrödinger equation to allow for relativistic effects, and had managed thereby to combine quantum theory and special relativity. The new Dirac equation led to a novel and seemingly crazy idea. To solve his equation, Dirac had to take a square root, which allowed two possible mathematical solutions: one with a positive sign, one with a negative sign. Dirac now postulated that the mathematics in fact spell out reality. In other words, there should be two types of electron: the old one, whose rest energy according to Einstein would be mc²; and the new one with energy —mc² (according to the solution of Dirac's equation).

Dirac even went a step further, proposing consequences for the empty region around a nucleus. Such a vacuum would really be a sea filled with negative-energy electrons. Here all possible states are occupied; electrons thus never jump from plus to minus energy because no more than one electron can occupy a given state, according to the Pauli exclusion principle. One could "see" such a negative-energy electron only by bringing the negative electron up to positive energies and by observing it and the hole it left behind as a pair of particles. This pair production can be achieved with the help of photons whose energy is high enough.

Reasoning in this manner, Max conjectured that the observed scattering was due to the negative-energy electrons in the vicinity of the nucleus. These are not free electrons, since all the states they could assume after a scattering event are occupied, according to Dirac's theory.

The intensity of this scattering turned out to be very difficult to calculate. Max struggled to arrive at estimations. With some advice from Hans Bethe, he got so far as to predict that this effect should vary proportionally with the fourth power of the nuclear charge. Here his contribution ended. Real progress in the calculations came twenty years later. In 1952, Hans Bethe and two of his graduate students at Cornell University published two papers in *Physical Review,* coining the term "Delbrück Scattering." By then Max had abandoned theoretical physics and held a position in the Biology Division at Caltech. He was astonished when his friends in physics told him about this curious reverberation from his former life.

The way Delbrück scattering is analyzed in modern quantum electrodynamics differs from the process Max suggested. His proposal that underground electrons caused the scattering of hard γ-rays was correct in principle but not applicable in this case. Experimentally, it turned out that the observed scattering component corresponded to pair production and annihilation. In other words, an incoming photon uses up its energy to lift an electron out of the Dirac sea to positive energies. The photon thus creates an electron-positron pair that in turn is influenced by the nuclear field. In the next step, the positron and an electron annihilate one another, producing two photons. In that way, light is scattered by a static electromagnetic field.

The Delbrück scattering problem relates to the question of how light scatters light. In the classical theory, two light beams just go right through each other without interacting. But in what is now called quantum electrodynamics, if one takes into account the negative-energy electrons, then the first light beam polarizes the vacuum, also scattering the second beam.

The effect Max had predicted—a contribution to the scattering by those underground electrons whose wave function is distorted in the neighborhood of the strong nuclear field—should also occur. The actual calculation of this effect and its experimental verification lingered for twenty years, because it turned out to be extremely difficult to test. In order to observe the effect, one must ascend to much higher energies than previously used, 10 MeV rather than 2.6 MeV. The existence of Delbrück scattering has since been confirmed experimentally.

LIFE UNDER THE NAZIS

Shortly after Max had begun working with Lise Meitner at the Kaiser Wilhelm Institute, the Nazis came to power and life in Germany changed. The Weimar Republic came to an end. The Weimar parliamentary system had never proven itself capable of giving effective government to Germany; the civil power lost its authority as soldiers and demagogues took over. The last free election of 1932 was won by Hitler's party, the National Socialists. When Reichspresident Hindenburg asked Hitler to form a government, the request enjoyed popular support, including that of Max Planck, who as president of the Kaiser Wilhelm Institute congratulated Hitler and pledged the fullest cooperation of the "Arbeiter der Stirn" (Workers of the Brain).

When Hitler became chancellor in January 1933, the international press did not seem to expect significant changes. But the Nazis acted quickly. They persuaded President Hindenburg that a Communist revolution was imminent, making him sign an emergency decree that suspended basic rights for citizens. Anyone could now be arrested on suspicion and imprisoned without trial. The Nazis turned the decree against their opponents by imprisoning them before the next election, scheduled for early March 1933. After an overwhelming victory, they introduced the "Enabling Act" *(Ermächtigungsgesetz)* that effectively made Hitler dictator of Germany, free from control by parliament or the president. By April 19, a new law threatened Jewish inhabitants of Germany. In the "Law for the Reestablishment of the Professional Civil Service *(Gesetz zur Wiederherstellung des Berufsbeamtentums)*, one paragraph ordered the "dismissal of civil servants of non-Aryan descent."

This caused the emigration of a large number of Max's colleagues, especially Jewish colleagues, leading also to harassment of those who did not leave. In his discussion of this period, Max often used the word "ridiculous"; he entirely disapproved of the Nazis. What angered him then was that the murky political situation made it difficult to concentrate on science. Max's boss, Lise Meitner, was half Jewish. At first, she was protected by her Austrian citizenship and by the fact that the Kaiser Wilhelm Institute was nominally a private institution; she thus was not a state employee. In addition, Max Planck, as the new president of the institute, powerfully protected many of his Jewish colleagues. But when Hitler annexed Austria in 1938, Lise Meitner could no longer be protected; she escaped to Stockholm.

The atmosphere of the period in Berlin can be felt from a story Max related about this period. At the neighboring institute, the Kaiser Wilhelm Institute for Physical Chemistry, Fritz Haber was the director. He was a man of great fame, having instituted the Haber-Bosch process of chemical nitrogen fixation, important for creating synthetic fertilizers. During World War I, Fritz Haber had invented chemical warfare. Though he was Jewish, his accomplishments at first buffered him from attack. Because his institute contained many Jewish associates, there was great concern about its future, as Max was told by his friend Karl Friedrich Bonhoeffer, who worked there.

Eventually, violent attacks on Fritz Haber appeared in the press; he decided to leave the country, and died a year later, in 1934. At first, there was no memorial service of any kind, but in 1935 the Kaiser Wilhelm Gesellschaft decided to stage a service in Dahlem, where most of the institutes were. That service became a bone of contention. The Nazi government tried to prevent it, forbidding any state employee (i.e., any professor) to attend. The principal speaker at the service was supposed to be Karl Friedrich Bonhoeffer, who in the meantime had joined the faculty in Leipzig. He arrived in

Berlin, only to receive a strict order from his superior, the Minister of Education, not to attend this memorial meeting. Finally Otto Hahn, co-director with Lise Meitner of the Kaiser Wilhelm Institute for Chemistry, agreed to read Bonhoeffer's speech. Karl Friedrich and Max circled the building, trying to decide whether they should attend the service. At last they determined that only Max should enter. From the back row, he watched Max Planck personally escort Otto Hahn to the rostrum to read Bonhoeffer's memorial statement.

STATISTICAL PHYSICS

In the early thirties, Max was extremely busy. In his job at the Kaiser Wilhelm Institute, he was performing tedious calculations on pair production by γ-rays; he was also working on nuclear models and had begun to help Lise Meitner write a book about the physics of the nucleus. In addition, he organized private seminars. In these times of confrontation, many Germans avoided large formal groups. A group of five or six theoretical physicists who felt they were in exile, though all were German, met regularly at Max's mother's home for private physics seminars planned by Max. No wonder members of the Delbrück clan in this period termed Max "a candle that burns at both ends."

In 1934, Max published his most comprehensive theoretical physics paper. In it, he attempted to understand the relation between statistical mechanics and thermodynamics from the point of view of the new quantum mechanics. Statistical and quantum mechanics have one concept in common: probability. The abstract formalism of quantum mechanics is in fact a device for making statistical predictions regarding the outcome of an experiment. Understanding probability seemed to provide the key to understanding physics.

In a joint effort with Gerd Molière, Max wrote what he called a "very learned" paper (D 11), by far the longest and most thorough of his theoretical papers; this was forwarded by Max von Laue to the Prussian Academy of Sciences. In forty-six pages, the authors presented an attempt to understand whether the paradoxes of irreversibility that had plagued classical statistical mechanics were mitigated by the uncertainty principle.

In classical statistical mechanics, the goal is to explain the increase in entropy. The Second Law of Thermodynamics states that in an isolated system, entropy can never decrease. Rudolf Clausins had introduced entropy in 1856 as a thermodynamical concept; it was later recognized by Ludwig Boltzmann as a measure of molecular disorder.

An isolated system evinces an obvious tendency to disorder. To explain this from first principles, one would have to utilize the laws that govern the motion of the individual particles which constitute the system under consideration (e.g., a volume of gas or a liquid). But these equations of motion embody the principle of time reversal; they are symmetric with respect to time, lacking direction.

The apparent paradox is that when one looks at individual particles, future and past seem to be interchangeable. Yet the whole system develops toward a more disordered state. The difference between past and future is created by statistical consideration. This becomes obvious if one understands the concept of probability as a prediction of a future event.

Mathematically, the problem was to extract from the theoretical foundations, which do not contain the arrow of time, a set of predictions that will include it. This is tricky. The problem that Gerd Molière and Max tried to tackle was to determine whether this paradoxical situation proved less paradoxical in a quantum mechanical formulation than in the classical one. After looking into the question very thoroughly, they found little difference. Superficially, one might have thought that the act of observation, which plays such a great role in quantum mechanics, introduced a directionality in time because one intervenes to change the future but not the past, thus creating an asymmetry in time. But it turned out that Max's attempt did not really cure the problems of statistical quantum mechanics.

Here Niels Bohr's strong influence on Max's thinking becomes manifest. Bohr had believed from the very beginning of quantum mechanics that in principle the concept of observation entails irreversibility. The entropy of a system must increase with every observation.

In Bohr's view, the arrow of time has to do with the shift from mechanical concepts like the motion of an individual particle to thermodynamical notions like temperature. This in turn provides another example of complementarity, as Bohr explained in his Faraday Lecture of 1932. Here he pointed out that "the very concept of temperature stands in an exclusive relation to a detailed description of the behavior of the atoms in the bodies concerned" (Bohr,

1932). As soon as one knows the temperature of a system, then—according to Bohr—the concept of energy has no meaning. Since determining the temperature involves an interaction between the object and a measuring device, the observational problem and thermodynamics considerations become complementary aspects of the whole description.

These issues of statistical physics remain difficult and still persist in scientific discussion. No wonder that Max was fascinated by the problems of quantum statistics all his life. He pondered the problem of time even in the late seventies, in a 1978 commencement speech at Caltech entitled "The Arrow of Time" (D 103). Here he pointed out that the directionality of time is the specialty of biology, where nothing makes sense except when viewed in a time frame from birth to death.

BEGINNINGS IN BIOPHYSICS

In 1934, Max was seeking a connection between biology and physics in a quite direct manner: he was looking into the influence of light on life. The discussion group in the Delbrück house in the Grunewald began to examine photosynthesis and photophysiology. Hans Gaffron was the photosynthesis expert. As a result of these discussions, he and Kurt Wohl published a series of papers on the kinetics of photosynthesis. An important piece of what could even then be called "molecular biology" came out of these publications and discussions. It can be described as follows: In photosynthesis, a plant has to make sugar out of CO_2 and water using light. Chemically speaking, the plant has to reduce CO_2. Theoretically, at least four quanta of light are needed to reduce one molecule of CO_2 to sugar. Experimentally, however, it was found that eight to ten quanta were required. The CO_2 molecule had to accumulate that many quanta of light before being metabolized to sugar and before the oxygen got liberated as molecular oxygen. After some sophisticated experiments, Gaffron and Wohl (1936) postulated essentially what is now accepted, namely, that photosynthesis proceeds in units consisting of about 1,000 molecules of chlorophyll, all funneling their energy into one photosynthetic reaction.

During this period, Max spent time investigating other biophysical topics. A former colleague from Copenhagen, Hendrik Kramers, invited Max in 1936 to give biophysics seminars in the Netherlands, leaving him free to choose any topic that interested him. Max picked population genetics and natural selection. His interest was stimulated by Ronald Fisher's book on the general theory of natural selection (Fisher, 1935). In Holland, as elsewhere, Max found physicists eager to learn about biology. Yet in the end his seminars had no effect whatsoever, for physicists did not follow him to biology then. After repeating this experience over and over, Max finally gave up proselytizing.

One of Max's seminars took place at the Phillips Laboratories in Eindhoven. When Otto Hahn heard about this invitation, he remembered that Phillips once made metallic uranium, just to see what it was like chemically. He speculated that Phillips would still have some sample bars left. No one else had metallic uranium at the time; scientists were using uranium salt (uranium nitrate). Otto Hahn asked Max to check whether Phillips could let him have a piece of metallic uranium. After the seminar, Max spoke to the director of research, who agreed and handed him a piece the size of a finger. Max put the metallic uranium in his pocket, returning with it to Germany. From the point of view of radioactivity, carrying the piece was quite harmless because the radiation was very weak. But the incident shows how informally such matters were handled in those days.

FRIENDS IN BERLIN

Hans Gaffron, Kurt Wohl, and their families shared a large house in Berlin. They often invited friends for musical evenings at which either the Wohls would play the piano or friends would perform string quartets, and Max regularly attended these gatherings. Here he noticed among the guests a small, inconspicuous, middle-aged woman in whom he became interested. Her name was Jeanne Mammen. As a painter, she had experienced an extremely creative phase in the 1920s, but the Nazis declared her expressionistic paintings degenerate. When Max met her in 1935, a life-long friendship began. At first, Max was a novice with respect to art; he felt uncomfortable with paintings that distorted reality. One day, when he was traveling with Jeanne on a tram to Wohl and Gaffron's house, he asked her why she painted people with black holes instead of eyes. Why would she adulterate nature? In answer, she pointed to the person sitting opposite them, saying, "Look, his eyes are just black and empty." After looking closely, Max had to agree (Figs. 13, 14).

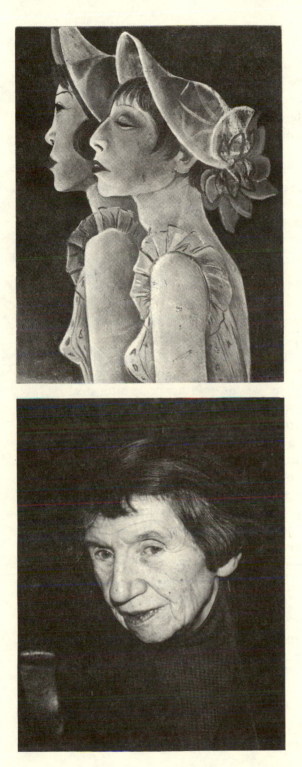

13. *Revue Girls, by Jeanne Mammen (1929). Courtesy of Berlinische Galerie, Berlin-West.*

14. *Portrait of Jeanne Mammen (1974). Photo by Gerd Ladewig.*

Under Jeanne's influence, Max began to take a serious interest in art. He carried some of her pictures with him when he emigrated to the United States, and organized a Jeanne Mammen exhibition at Caltech in 1938. Later, his Pasadena home and his Caltech office were filled with her paintings.

Around the time Max met Jeanne Mammen, he was also introduced to Martha Dodd, the daughter of the American Ambassador in Berlin. They became close friends. At first, Dodd was the darling of the Nazis, but she soon became a friend of the Soviet Union. In 1957, American newspapers described her as a "red agent." Today, her record has been cleared.*

HABILITATION AND INDOCTRINATION

While Max was devoting so much time to biology, he was still Lise Meitner's assistant in physics, his job being to provide theoretical explanations for the experimental data. He was not paid for his efforts in biology. Like all others who wanted an academic career in Germany, Max had to prove his qualifications after his first few years of work, a process known as habilitation. In Germany, that meant presenting a paper, called "Habilitationsschrift," in which one reviewed a field and described one's own contributions to it. Since Max's degree and position lay in physics, he picked Dirac's theory of the positive electrons. In great detail Max outlined his ideas about the scattering of light on heavy nuclei such as lead. He entitled his habilitation "Beiträge zu Diracs Theorie des positiven Elektrons" (Contributions to Dirac's Theory of the Positive Electron). No one seems to have retained a copy.

In May 1934, Max sent this essay to the dean of the Philosophical Faculty of the Friedrich Wilhelm Universität in Berlin, with an application for habilitation. If accepted, he would be called Privatdozent (a lecturer) and could then teach at a university. Although Privatdozent cannot really be termed a job, because it involves no pay, it was a prerequisite to applying for a tenured professorship.

Max's habilitation was quickly turned down without any explanation. The likely reason was his failure to pass muster politically. The Nazis had added this new twist to the whole procedure of

*The indictment charging Martha Dodd with conspiring to spy for the Soviet Union was finally dismissed in 1979 (*New York Times,* March 23, 1979).

becoming a permanent member of a university. In addition to scientific qualifications, one had to demonstrate political reliability. Indeed, the Nazis forced politics into every private corner, trying to control what people were thinking, and using terror in order to enforce agreement with their views. This invasion of private life caused an unbearable situation for Max. He no longer felt at home in Germany, but in 1934, he had nowhere to go.

Max tried to follow orders. Since he wanted to become a lecturer at the university, he had to fulfill the Nazis' condition: to gain approval politically. To do so, he had to attend a Dozentenakademie, as it was called; one can best describe it as an indoctrination camp. About thirty people would gather in discussion groups to hear lectures on the new politics and the new state. After three weeks of "free" discussions, the organizer determined whether a participant was politically mature enough to lecture at the university.

When Max joined an indoctrination camp in 1934 at a site opposite Kiel, on the Baltic Sea, he found the experience fascinating at first. For three weeks, the invited scientists lived on a lovely estate with a large park, three to a room. The marvelous thing for Max was that here, for the first time in his life, he was thrown together with people from other disciplines. He shared his room with an economist and a psychiatrist, thus learning much about these other sciences. The main activity, of course, was to listen to enthusiastic lectures by reliable party members; the participants were extremely nervous, not knowing what could and couldn't be said. Max was too outspoken; he was informed afterwards that he wasn't quite mature enough.

He had obviously shown he thought that Nazis were fools. Max liked to tease them. He needled them about their race theories and refused to join in singing the Nazi hymn, "Horst Wessel Lied," claiming he found two lines unclear:

> *Kameradan, die Rot Front und Reaktion Erschossen*
> *Marschieren im Geist in Unseren Reihen mit*

> (The spirit of comrades [who] shot Reds
> And Reactionaries marched along with us in our ranks)

> or

> (The spirit of comrades shot [by] Reds
> And Reactionaries marches along with us in our ranks)

These German lines leave doubt as to whether the comrades have shot or been shot. The Nazis, not open to subtle arguments, sent Max home.

In June 1935, he tried again. This time he faced an additional problem typical of the daily difficulties common to life in Nazi Germany. Max's dilemma involved how to sign a letter to the Nazis. By that time, it was conventional to sign with "Heil Hitler." In his first application, he had closed by substituting "With my great respect" as a complimentary closing. However, if he now refused to precede his signature with the obligatory "Heil Hitler," he would have no chance of passing the political test. One of his friends solved the problem for him by pointing out that "With my great respect" involved just as big a lie as "Heil Hitler," so Max signed the letter in the desired fashion. Clearly, all those who stayed in Germany had to make such compromises frequently.

Nevertheless, Max also came up short in his second stay in indoctrination camp. It must have been transparent that he was not greatly enamored with the new regime. The Nazis produced a strange compromise, giving him permission to use the title Dr. Phil. Habil. in August 1936, but not allowing him to lecture at a university. He was simply permitted to continue working in the Kaiser Wilhelm Institute.

It became clear to Max that a university career in Germany was unlikely for him. He had been turned down despite his considerable efforts to prove that he was not Jewish. This involved supplying authentic copies of all the baptismal certificates of his four grandparents and their Christian marriage certificates. Max had to turn in such documentation even from his great-grandparents, none of whom was Jewish.

The situation was saved by a second Rockefeller Fellowship, which enabled Max to go to Pasadena in 1937.

THE MISSING OF FISSION

By moving to the United States, Max missed taking part in an important discovery in 1938, the discovery of nuclear fission by Lise Meitner. Max in his later years never deplored this misfortune. On the contrary, he pointed out that his departure proved a necessary requisite for the discovery. With typical honesty, Max felt that he had prevented Otto Hahn and Lise Meitner from an earlier

breakthrough, since he had not really been concentrating on physics and had misinterpreted some experiments.

In the 1930s, Otto Hahn and Lise Meitner were considered the
experts on radioactivity and on the chemistry of radioactive substances. They had followed up on a discovery by Enrico Fermi that
by irradiating a large number of chemical elements with neutrons,
one could obtain radioactive substances. Fermi found especially
that irradiating uranium with neutrons yielded quite a number of
such substances. Fermi suspected they were elements with an order
number larger than 92; he called them transuraniums. Hahn and
Meitner discovered that upon irradiation of uranium by neutrons,
a large number of products arose which could be characterized by
their half-lives and by the type of radiation they gave off. Hahn and
Meitner interpreted these products to be elements with order numbers 93 to 97. That was in 1935. Fermi's discovery seemed to have
produced some excitement also in Max, though he later claimed to
have had only limited interest in the nuclear reactions. In his autobiography (Hahn, 1968), Otto Hahn describes Max's astonish-
ment that Hahn and Meitner, "after the exciting news from Italy,
could peacefully go to sleep without having repeated the experiments."

So Max was involved, but the situation became more and more
confusing when, in the following year, Hahn and Meitner detected
quite a few more "elements," which were all thought to be isomers
of these transuraniums. Without taking too much interest (because
it wasn't biophysics and because this was not really a theoretical
physics problem, seeming too trivial for that), Max was very quick
in interpreting all of these as isomers supposedly created by the
neutrons. Since this explanation was accepted, Max successfully—
as he phrased it—"prevented the early discovery of what was
really going on, namely, fission" (I). Otto Hahn and Fritz Strassmann did not find this out until 1938, one year after Max had
left.

In 1936, no one saw that fission of the uranium was occurring.
Yet it was something that any physicist could easily have figured
out. One did not need any calculations; all one needed to know was
that there was excess energy there, something Einstein's famous
formula $E = mc^2$ would indicate. Some mass of the nucleus could
be transformed to energy. Thus when the neutron enters the nucleus, there is enough energy around to blow it to pieces. One

needed only to be able to add and subtract. Fission just did not occur to physicists until they were literally forced to this conclusion in 1938. Max himself had by then arrived in Pasadena and given a seminar at Caltech's Physics Division on his interpretation of the elements as isomers. His view was proven wrong a few weeks later.

Earlier in the same year, Max had spoken about the new developments in nuclear physics at a seminar in Stettin. He had been asked to replace Hahn as the speaker, since Hahn had been unable to come. During the discussion following Max's speech, one listener inquired if nuclear physics might prove of any economic importance. Max replied that this would be like washing dishes with champagne. It would work, but nuclear processes are too expensive and not really an improvement.

The path toward the discovery of fission was forged when Irene Curie in Paris reported finding thorium as a decay product (Frisch, 1979). This would imply that a uranium nucleus, on being hit by a neutron, could emit an α-particle (a helium nucleus). It seemed unlikely. But when Hahn and Strassmann repeated the experiment, they found even smaller products, such as radium, no less than four places below uranium (order number 88 compared to 92). When they communicated these results to Lise Meitner, she expressed strong doubts and demanded irrefutable evidence. By that time, Meitner had been forced to leave Germany and was living in Sweden.

Hahn and Strassmann now embarked on a series of careful tests. To their utter surprise, they found barium (order number 56) and crypton (order number 36). They published these findings with great reluctance (Hahn and Strassmann, 1939), stating: "as chemists we have to say that the new product is not radium, but it is barium, since elements other than radium and barium are out of the question." In a subsequent publication, they did go a step further to interpret their finding as a split of the uranium atom.

Otto Hahn had sent a letter to Lise Meitner, whose nephew Otto Frisch was then visiting her. Frisch had come up from Copenhagen, where he worked in the Niels Bohr Institute. Both tried to imagine where the energy to break the nucleus apart might have originated (Frisch, 1979). They noticed that the two nuclei that formed after the impact of the neutron together possessed slightly less mass than the nucleus of the original uranium. But according to $E = mc^2$, much energy exists in a little mass. Meitner and Frisch had found the source for the energy of the product nuclei.

Back in Copenhagen, Frisch asked an American biologist the correct term for the division of individual cells. The answer was "fission." When their explanation was published, Frisch and Meitner coined the expression "nuclear fission" *(Kernspaltung)* (Frisch and Meitner, 1939).

Light and Life—
The Great Challenge

THE IMPACT OF A LECTURE

A lecture by Niels Bohr at 10:00 A.M. on August 15, 1932, changed the course of Max's life. "It was the opening lecture of an International Congress of Light Therapists," taking place rather solemnly in the Rigsdag in Copenhagen, in the presence of "[the Danish] Crown Prince," the prime minister, and "many distinguished looking gentlemen from all parts of the world" (D 95). Max had arrived that morning by the night train from Berlin, and was met at the station by Léon Rosenfeld, who told him immediately that Bohr wanted him to attend the lecture.

Copenhagen hosted many international meetings, and Niels Bohr was quite often asked to dignify the opening sessions with a lecture. He liked to present talks, although from the point of view of information transfer, his lectures were invariably dismal failures. Bohr always lectured without manuscript or notes, whispering in English with a strong Danish accent. He always picked the same topic: the possibility that the recent successes in clarifying the meaning of quantum mechanics might have wider implications for other sciences.

On August 15, 1932, the situation was aggravated by the fact that practically no one in the audience knew anything about quantum mechanics; also "the immediate effect of the lecture was somewhat hilarious due to a comic mishap" (D 95). When Bohr was called upon to speak, he promptly got lost behind the rostrum. He finally found the lectern, only to whisper almost inaudibly. Subsequently, he must have actuated a switch that caused a hydraulic mechanism to lift the lectern: Bohr gradually disappeared behind it. The audience watched the lectern in utter fascination. At last Bohr pressed the switch down and continued. From then on, of course, every member of the audience riveted his attention on Bohr, anticipating a repeat of this mishap.

This was the great lecture "Light and Life," published a year later (Bohr, 1933); it had a decisive influence on Max's professional thinking and career. In "Light and Life," Bohr proposed "the bold idea that life might not be reducible to atomic physics." He suggested that "there might be a complementarity relation between life and atomic physics analogous to the complementarity encountered with the wave and particle aspects in atomic physics. The result would be a kind of uncertainty principle regarding life, analogous to Werner Heisenberg's uncertainty principle in quantum mechanics" (D 95).

The actual lecture, which differed from its published version, made an enormous impression on Max because in it Bohr went out on a limb to predict such a complementarity; here he was spelling out an explicit theory that later on could be proven wrong. Usually Bohr was very careful never to say anything that could turn out to be "wrong," being very cautious in his formulations. He liked to warn his audience that every one of his sentences must be understood not as an affirmation but as a question. This makes it difficult to try to paraphrase Bohr. Max felt that every quote from Bohr should be introduced as a questioning statement.

In the 1932 lecture, though, Bohr positively concluded that the processes of life are complementary to physics and chemistry. This certainly challenged Max to take the argument seriously and to follow it up. He would do so for the remainder of his life.

THE NOTION OF COMPLEMENTARITY

To discuss Bohr's interpretation of quantum mechanics is dangerous. On the one hand, one cannot quote his lectures, for reasons described above. On the other hand, his published works require agonies "to get out of them what he put in" (D 95). Bohr quite often complained that neither his supporters nor his opponents understood or described the notion of complementarity properly; all have distorted it. Bohr once even scolded Max for this.

Bohr's notion of complementarity was meant to reconcile opposing points of view. Unfortunately, the idea itself split the scientific community in half. Some think Bohr's argument is more important than all of quantum mechanics and that it eventually will reveal deep insights if only taken seriously (Primas, 1981; Stent, 1979; Jauch, 1973). Others find it trivial at best, or simply unnecessary. To Bohr, the complementarity was obvious; but Einstein rejected it, arguing that since a clear-cut definition was impossible, the notion proved of no value. Indeed, when Bohr introduced the term "complementarity" in 1927, he did not define it—he just used it. Here Bohr referred to the space-time picture and the demands of causality as "complementary but mutually exclusive" features of the way experience is described. Bohr emphasized that complementarity is meant to indicate completeness; it unites conflicting aspects that belong together.

In atomic physics, experimental observations are collected by essentially irreversible processes of amplification. Observations become inseparable from their respective experimental set-ups. Bohr's complementarity argument recognizes that experimental situations used to define observations may prove mutually exclusive. The quantum mechanical formalism permits one to derive predictions for the outcome of a certain experiment from the results of experiments made with a mutually exclusive arrangement, if they are done successively; these predictions will be purely of a statistical and probabilistic nature.

Bohr immediately realized that complementarity is related to a general difficulty in the formation of human ideas, since the notion applies to situations where it becomes essential to distinguish between subject and object. Bohr noticed that the answer to the question "What is light?" depended on the observer. If one wanted to know if light was a wave phenomenon, one could perform an interference experiment and receive a positive answer. If one wanted to know if light showed particle behavior, one could study scattering phenomena and also receive a positive answer. The same was true, as we have seen, in work with electrons.

This dualistic nature of atomic phenomena produced confusion, which Bohr tried to clarify with the complementarity argument. He proposed that it is wrong to term electrons as "particles" and to maintain that only unfortunate circumstances prevent the detection of their orbit in a hydrogen atom. He pointed out that the atomic reality might simply be too rich to be caught in the net of the concepts that classical physics provides, suggesting that only complementary descriptions like "wave" and "particle" would provide a full account of the atomic phenomena. The necessity underlying the complementary approach is the simple fact that experiments with atomic particles are performed with macroscopic instruments, and the results have to be communicated in everyday language; such language in turn is adapted to macroscopic phenomena and uses the concepts of classical physics.

Complementarity is a philosophical issue, and as such it faces a difficult time in most sciences. In physics, however, the concept was rapidly accepted as relevant, since at the time Bohr introduced the term Werner Heisenberg described a related phenomena. Heisenberg demonstrated that for an electron, one cannot simultaneously determine an exact position and a precise momentum. He discovered the principle through pondering whether one could build a microscope to see an electron. He realized that in order to locate the electron's position, one had to use light waves of very small wavelength. But those waves, having a high energy, would scatter the electron and thus prevent the determination of its speed. Heisenberg concluded that quantum effects do not permit an experiment which can determine the exact position and momentum of an atomic particle at the same time. If one parameter is measured, the other fluctuates and cannot be pinpointed exactly. Heisenberg's uncertainty relations can be interpreted as the quantitative basis of the complementarity argument.

BOHR'S GENERAL BIOLOGICAL ARGUMENT

The fundamental and decisive aspect of quantum theory that is relevant to Bohr's argument is the fact that there are experimental arrangements which are mutually exclusive. In biology, one is dealing with systems of such intricate complexity that it is impossible to make the observations necessary for a proper quantum mechanical treatment. A set-up to gather such observations would be mutually exclusive with a situation in which the proper phenomena of life are manifest. It is therefore reasonable to expect that a full development of biology necessarily involves concepts which are not reducible to those of physics.

This is roughly the argument that Niels Bohr presented in his "Light and Life" lecture in 1932. The crucial passage in the printed version reads as follows:

Thus, we should doubtless kill an animal if we tried to carry the investigation of its organs so far that we could describe the role played by single atoms in vital functions. In every experiment on living organisms, there must remain an uncertainty as regards the physical conditions to which they are subjected, and the idea suggests itself that the minimal freedom we must allow the organism in this respect is just large enough to permit it, so to say, to hide its ultimate secrets from us. (Bohr, 1933)

This is not what Max remembered having heard in the lecture. In the talk, according to Max's memoirs, Bohr used the following analogy:

In physics it is obvious that even in the simplest case such as a proton running around an electron one can do classical physics until one's dying day and never get a hydrogen out of it. In order to achieve this, one has to use the complementarity approach. If one looks at even the simplest kind of cell, one knows it consists of the usual elements of organic chemistry and otherwise obeys the laws of physics. One can analyze any number of compounds in it but one will never get a living bacterium out of it, unless one introduces totally new and complementary points of view. (A)

Max was intrigued with the idea that complementarity could reveal great simplicity in a hopelessly complex situation. He was fascinated by this notion the moment Bohr presented it. As a first consequence, it motivated him to scrutinize the writings of Immanuel Kant on causality to see how the German philosopher could have overlooked this epistemological possibility; Max found this situation utterly removed from anything that Kant had conceived. For Max, there was no doubt that physicists had been pushed in epistemological directions not dreamed of before.

MAX'S BIOLOGICAL MOTIVATION

The complementarity argument in biology has been described in such detail because of its importance to Max's thinking. As Max wrote when he invited Niels Bohr in 1962 to deliver a lecture in Cologne, "There is no complex of scientific questions that affects me more deeply and it has through the years provided the sole motivation for my work." Max struggled with the argument for the rest of his life, always hoping that complementarity in biology might be just around the corner. It seemed so obvious to Max in 1932:

> The mysteries of life, in those days, from the point of view of physics, were indeed stark: cell physiologists had discovered innumerable ways in which cells responded "intelligently" to influences from the environment, and embryologists had demonstrated such feats as each half of an embryo developing into a complete animal. Such findings were vaguely reminiscent of the "wholeness" of the atom, of the stability of the stationary states. (D 95)

What really fascinated him was that "the stability of the gene and the algebra of genetics suggested something akin to quantum mechanics" (D 95).

Max's understanding of Bohr's general argument can be stated as follows: Observations are defined by experimental situations. Some of these observations might not be expressible in terms used in classical physics. After all, biologists talk about genetics in terms of information content and of concepts of potency and determination. Since some experimental situations are complementary to each other, there is always a chance that certain theories attempting to achieve an interpretation of the observations will involve a mutually exclusive argument between a physical quantity and a nonphysical concept. That would be a higher form of complementarity, different from the purely physical one. The practical way to investigate Bohr's argument is to find a manifestation of life that cannot rationally be reduced to an interpretation in molecular terms. In other words, looking for complementarity does not alter one's approach to biology. The avenue of progress is to develop better techniques and to conduct more refined experiments, as biology has traditionally been conducted.

In Max's case, waiting for complementarity meant the following research program: He would investigate a biological system "as something analogous to a gadget of physics," with the "hope that when its analysis is carried sufficiently far, [it] will lead to a paradoxical situation analogous to that into which classical physics ran in its attempts to analyze atomic phenomena," as he wrote to Bohr in December 1954. Max wanted to achieve in biology what Bohr and Rutherford had done in physics in the second decade of the century.

Max's friends will note that he hardly spoke to them about complementarity. Some of his best friends may never even have heard him deal with this concept. While it is true that Max talked a lot with colleagues, he could also keep things to himself. This was especially true if he was deeply moved by something.

"UNFINISHED BUSINESS"

To use a term not yet coined in 1932, Max's hope was that complementarity might be encountered during the construction of a "molecular biology." This term refers to attempts to reconcile elementary biological phenomena—such as nerve conduction, transport through cell membranes, excitation in sensory elements, and replication of genetic material—with molecular pictures. Max's dream has not come true, as he pointed out himself in 1981:

> Indeed, we might say that the discovery of the Double Helix in 1952 did for biology what many physicists had longed for in atomic physics: a resolution of all the miracles in terms of classical mechanical models, not requiring an abdication of our customary intuitive expectations. The Double Helix, indeed! With one blow the mystery of gene replication was revealed as a ludicrously simple trick, making those who had expected a deep solution feel as silly as one might feel when shown the embarrassingly simple solution to a chess problem one may have struggled with in vain for a long time. The Double Helix, indeed! It does not matter that the mechanics of replication of nucleic acid has turned out to be enormously more complex than was thought in the first flush of victory and that even now vast uncertainties remain in the most basic area of molecular biology. Never mind! We now understand that organisms can be viewed very successfully as molecular systems, of enormous complexity, to be sure; nevertheless, an upper limit can be set to the complexity and even more powerful methods to probe it are being developed at a mind-boggling rate. (D 95)

That is, the double helix did show life as reducible to classical principles. No contradiction showed up.

Complementarity now seems a dead issue; molecular genetics does not need it. Max, on the contrary, considered it "unfinished business." While it might not be needed in genetics, it might still be involved in other areas of biology.

Aside from its resolution of paradox, there was yet another reason why the principle of complementarity made such an impression on Max. He hoped to find a simple solution in biology in which the traditional approach appeared to bring higher degrees of complexity. Max yearned to discover simplicity behind the complications.

MAX'S VERSION OF THE ARGUMENT

Max was not the only one of Bohr's pupils who took the notion of complementarity seriously. Pascual Jordan also tried to apply it to biological questions. While Max eventually did so experimentally, Jordan adopted Bohr's idea as a method to consider biological problems theoretically. Max did not approve of Jordan's efforts. He complained to Bohr that Jordan's abstract lectures on the implications of the complementarity argument would only confuse others who might be interested in the topic.

At one point, Max decided to develop his own formulation of the complementarity argument that could prove understandable to biologists. In a rather derogatory fashion, he explained in a letter in German to Niels Bohr on November 30, 1934, that a short version of the complementarity argument was necessary because "biologists tend to read comprehensive papers in a superficial way. That's why they never will be able to grasp a new subtle idea properly." In this letter, Max then presented his statement for biologists, one enthusiastically greeted by Bohr.*

Our claim: Those assumptions that are necessary for the existence of causality in biological phenomena may partly contradict the laws of physics and chemistry, because experiments with living organisms are certainly complementary to those that determine the physical and chemical events on the atomic level.
Explanation: (1) We do *not* state that the laws of atomic physics can explain the *specific phenomena* of life. On the contrary!
(a) The laws of atomic physics are the common root of physics and chemistry. It was believed, in the old days, that chemistry could be reduced to classical physics. Now it has been shown that one can and has to introduce formally contradictory assumptions, simply because the common root is at the atomic level and because the experiments in this field are complementary in nature. E.g., the experiment in which I

*The original is in German; the following translation is by E.P.F.

synthesize a chemical compound (a macroscopic experiment) is complementary to an experiment for determining the orbit of those electrons that produce the chemical bond.

The justification that was developed afterwards to account for the contradictions (quantum jumps) arose from the insight that a description in atomic detail can only refer to an atomic experiment, in which the experiment—as one says—disturbs the phenomenon. More accurately: An experiment, in which the observer and the observed (the object) cannot be distinguished uniquely. Thereby the possibility for a causal description vanishes.

(b) Since obviously in living organisms chemical and physical events are connected down to the atomic level, the common root of biology *and* physics *and* chemistry *has* also *to be* on the atomic level. Therefore a causal description of the connections *cannot* be achieved with physical and chemical notions alone. For on the atomic level physics and chemistry do not permit a causal description.

(2) We do *not* state that a biologist in his experiments does kill life or that he has to kill it.

On the contrary!

In genetics *and* in development *and* in psychology *and* in biochemistry *and* in biophysics it is characteristic, *and* essential, to investigate processes in the living organism. Because of that, these methods of research cannot gain results about the individual atomic elementary processes. They are far away from it, as everybody agrees here. They even have to stay away from the atomic level for their description to remain *strictly* causal. For that reason, these areas of research are not causally reducible to each other, just as physics and chemistry are not reducible in that way.

Originally, Max had asked Bohr to publish a short, concise version of the complementary argument intended for use by nonphysicists, omitting discussions of the mathematics of the uncertainty relations. But his hopes were dim, because final editing for Bohr always took very long, even though Max had tried to get things moving.

THE GENE IS THE THING

Quantum action had been discovered by Max Planck when he examined the thermodynamic equilibrium between radiation and matter. Max began his search for complementarity in biology by analyzing how ionizing radiation influenced the genetic material. Genes were stable elements; as a special feature, they could be shifted to a different form, which again was stable. Could the new quantum mechanics explain this, or did biology run into a paradox right here?

Although genes definitely belong to biology, their properties have always attracted the attention of physicists. In fact, they were discovered by an Austrian monk who had been trained as a physics teacher. Gregor Mendel (1822–1884), like Max, failed his examination, ending up in a quiet garden in Brünn where he founded genetics by establishing the laws of inheritance. Observing garden peas, Mendel noticed patterns in the way progeny inherited properties in which parents differed. Through such observations, he demonstrated the existence of "elements of inheritance" *(Erb-elementé)*, which were termed "genes" in 1909, the same year the electron was named. In a sense, Mendel discovered the atoms of biology, the indivisible units that formed the basis of what one could see; but in the second half of the nineteenth century, there was little faith in the existence of entities such as atoms. The notion of atoms and molecules did exist to interpret the rules of chemistry, but it did not seem they would ever "materialize." Only at the turn of the century did atoms turn out to be real and of definite size and structure. Mendel's results preceded this, however, and few took his work seriously.

At the beginning of the twentieth century, when Max Planck discovered fundamental discretenesses in nature, Mendel's papers were rediscovered. In particular, H. de Vries's observation of "jumping" variations attracted biologists' attention (de Vries, 1901). Such mutations did not occur in any intermediate form.

Quantum physics went on to develop a powerful theory to explain the stability of atoms, the occurrence of chemical bonds, the structure of matter, and its interaction with light. In genetics—the name was used for the first time in the year Max was born—the chromosomal basis of inheritance was discovered. By working mainly either with the fruit fly *Drosophila* or with maize, geneticists detected with the light microscope structures called chromosomes that followed the pattern of inheritance described by Mendel's laws. The 1930s have been described as the "climactic" decade of classical genetics (Dunn, 1965), in which chromosomes were observed and their behavior analyzed for the first time. But genes remained invisible and thus abstract constructions. What was their nature, and what was the mechanism of their mutation?

In 1927, Herman J. Muller announced that he had induced mutations in flies with X-rays. This discovery opened the way for physicists to look for physicochemical processes in the biological entity called the "gene." For this breakthrough, Muller was awarded the 1946 Nobel Prize for Medicine. He later succeeded in locating the position of genes on the chromosome. Muller postulated the gene to be the true "basis of life," realizing that "the geneticist himself is helpless to analyze . . . [its] properties" (Muller, 1936).

Muller urged the use of modern physics for analyzing the gene. In 1932, he came to Berlin to extend the study of comparative mutagenesis with a Russian émigré, Nicolai Timoféeff-Ressovsky, who ran the genetics laboratory at the Kaiser Wilhelm Institute for Brain Research in Berlin-Buch. Together, they tried to provide definite proof that ultraviolet radiation, like X-rays, caused mutations (Carlson, 1981).

Muller and Timoféeff-Ressovsky studied the types of genetic changes induced in *Drosophila* by different wavelengths of radiation. They found that they could make exact quantitative studies of the frequency of mutations obtained at different doses. With this data, they tried to explore the mechanisms by which physical events change the individual gene or break chromosomes. Muller was particularly interested in meeting physicists and recommended that Timoféeff do so too. He was convinced that a multidisciplinary approach to genetics would yield far greater insights into the gene and the nature of life than genetics alone could accomplish.

Muller encouraged such collaboration in spite of a disappointing meeting with Niels Bohr. After Muller had lectured on the gene in Copenhagen in the spring of 1933 and held discussions with the Danish physicist, he concluded that Bohr's "ideas in biology were hopelessly vitalistic" (Carlson, 1981).

At the time Muller and Timoféeff were collaborating in Berlin, Max had begun to organize the private discussion group in his mother's house. One of the people Max brought into those seminars was Timoféeff, who attended these meetings even though his institute was at the other end of Berlin and it took him an hour and a half by various public conveyances to get to Max's home.

The discussions during these gatherings were fairly intense, according to Zimmer: "we talked for ten hours or more without any break, taking some food during the session" (1966). The idea of working in groups had always seemed quite natural to Max. Today, teamwork is normal in science, but in Germany in the thirties it was the exception; "interdisciplinary team-work appeared rather strange to some scientists" (Zimmer, 1966). Max, though, had spent all of his physics life in groups, where he saw that teamwork was successful. He had seen firsthand the productive collaboration between Hahn and Meitner, and knew that in the development of genetics, the truly influential research had been conducted by a team consisting of T. H. Morgan, C. B. Bridges, H. J. Muller, and A. H. Sturtevant at Columbia University.

Max's purpose was to see what the new ideas in physics could contribute to the problems of biology; he thus asked biologists and biochemists to join the discussion in the group. When Max contacted Timoféeff, the Russian geneticist was ready not only to show his fruit flies to the physicist but also to discuss mutation research with him. Muller and Timoféeff's joint work in Berlin had provided evidence that individual ionizations were responsible for gene mutations. The prospect of applying modern physics to these biological phenomena proved irresistible, and shortly after Muller left Berlin, Max began a collaboration with Timoféeff-Ressovky and K. G. Zimmer that led to a quantum model of gene mutation. The theory turned out to be incorrect, but it was "a successful failure" (Carlson, 1966); the stimulation it provided was important. The analysis the three men offered in 1935 (D 8) constitutes one of the earliest approaches to molecular biology.

Through the collaboration of Max's trio, genes turned out to be understandable in terms of physics. At the time of its discovery in 1935, this result was by no means trivial. We can see this from a 1939 quotation, when Caltech's two geneticists—A. H. Sturtevant and G. W. Beadle—published their *Introduction to Genetics.* In the preface, they described their splendid isolation:

> Physics, chemistry, astronomy, and physiology all deal with atoms, molecules, electrons, centimeters, seconds, grams—their measuring systems are all reducible to these common units. Genetics has none of these as a recognizable component in its fundamental units, yet it is a mathematically formulated subject that is logically complete and self-contained. (Sturtevant and Beadle, 1939)

Despite establishing a firm foundation for molecular genetics, the victory of quantum mechanics in the 1935 quantum model of the gene meant the first damper to Max's hope. For complementarity was not needed to understand the stability of genes. No paradox emerged from this study; no new epistemological position was required to understand genes; and no new laws of physics were uncovered.

THE NATURE OF THE GENE

The "three-man-paper" was published in 1935 under the title "About the Nature of the Gene Mutation and the Gene Structure." This teamwork immediately demonstrated the value of a joint effort. Timoféeff directed the activity, for the group was pursuing his line of research. In order to study quantitatively the induction of mutations by ionizing radiation, one had to have someone who could analyze radiation doses quantitatively. K. G. Zimmer was the physicist responsible for that. One needed a dose-response curve, that is, a graph giving the relationship between the number of mutations and the radiation dose. Timoféeff provided the ordinate, while Zimmer produced the abscissa. Finally, Max developed a theory and worked out a quantum mechanical model of the gene. Three scientists with different backgrounds were necessary for the enterprise.

As a main result, the gene could now be connected with the measuring systems of physics and chemistry. It changed from an abstract unit without dimension to a macromolecule. The German geneticist Peter Starlinger has described this achievement as "a scientific revolution." By combining different fields of research, the group reached an insight that surprised geneticists, though not physicists. The revolution was accomplished simply by interdisciplinary collaboration.

In their paper, Timoféeff, Zimmer, and Max also suggested that genes would prove to be "the ultimate units of life" *(letzte Lebenseinheiten)*. They thus anticipated the analysis and manipulation of living organisms at the level of the gene.

GENES AS TARGETS

In the nineteenth century, Mendel not only postulated that there were genes but also introduced the distinction between a recessive and a dominant variant. A body cell of a higher organism has two copies of each gene, which are usually identical. If a mutation in only one copy can lead to an observable trait, it is called a dominant mutation. A recessive mutation requires two mutated genes to manifest itself.

Genes became susceptible to physical analysis once it was established that Mendelian elements were related to visible structures (chromosomes) inside the cell. The first evidence in favor of such a connection was presented in 1902 by the American W. Sutton. Working with grasshoppers, he demonstrated compelling parallels between chromosomal and Mendelian factor transmission. The physical basis of Mendelian genetics was firmly established in the "Fly Room" at Columbia University by T. H. Morgan and his co-workers, A. H. Sturtevant, C. B. Bridges, and H. J. Muller. Their work analyzed *Drosophila,* which offered a new generation every ten days and beautifully visible traits such as eye color and wing shape.

By studying the linkage of certain traits with the sex of the fly, they could establish that in *Drosophila,* sex could be attributed to certain chromosomes, called X and Y. The male fly has one of each, while the female fly has two X-chromosomes. This sex linkage in *Drosophila* was not the only discovery made in the "Fly Room"; the chromosome theory of Mendelian inheritance was also worked out there, including the demonstration that genes are material substances that can be mapped on chromosomes. The resulting book, *The Mechanism of Mendelian Heredity,* by Morgan, Sturtevant, Muller, and Bridges (1915), is of historic importance.

Several years earlier, Muller had been the first to notice that with an increasing dosage of X-rays, the number of mutations would also increase. (The immediate effect of these X-rays is to create ion pairs from water molecules: the ions in turn act on the genetic material.) The three men in Berlin reproduced this result to find that the number of recessive mutations in the X-chromosome of *Drosophila* was proportional to the radiation dose, if one measures this dose in terms of ion pairs produced or in terms of small clusters of ion pairs. This occurred for both hard and soft X-rays, at high as well as low intensities; it occurred even in the case of gamma-rays. It was also true whether or not the dose was fractionated.

By a more detailed analysis, the group could even conclude that a single hit sufficed to produce a mutation in a gene, with a hit being defined as an ionization in some cellular structure. This could all fit together in a clear-cut picture. Genes were units that could be affected by radiation; they were targets. With the dose-response curves available, one should be able to calculate the size of the target (i.e., the size of the gene) and figure out the energy that was required to induce a mutation. Since a single hit sufficed to produce a gene mutation, it should be possible to arrive at an estimate of the gene stability.

The mathematics used for such an effort is called target theory. In it, one tries to compute the probability that a given structure in the cell will actually be affected by the incident radiation. One tries to calculate the "sensitive volume" of the object that is apparently hit by the ionizing radiation. This method was first applied to a biological problem in 1924 in an investigation of the influence of X-rays on chick embryos growing in vitro. The target in this case was calculated to be of the order of one thousandth of a millimeter, considerably smaller than the diameter of the cells, which were one hundredth of a millimeter.

Max based his calculations of a sensitive volume on a study of the region of the *Drosophila* chromosome that is responsible for the white color of the flies' eyes. From knowing which dose would produce an average of one mutation in a given number of cells containing only one copy of a gene, he calculated a volume of the target sphere, arriving at a length of 9μ for his gene.

Today, the average length of a gene can be calculated from what is known about the double helical structure of the genetic material and the size of proteins. It is now known that $1\ \mu$ DNA holds an average of 6.5 genes; thus Max's gene encompassed about 60 "modern" genes.

THE STABILITY OF THE GENE

The most important conclusion that the three men could draw from their study was that the interaction of radiation with the genetic material—a hit in a target—caused a "quantum jump" which made itself visible as a mutation. In other words, the stability of a gene (and of its changed form, the mutation) could be explained in a similar fashion to the stability of a regular molecule—in terms of quantum physics.

The stability of atoms had been explained by recognizing that they have only discrete series of states to choose from. The change

meet until 1941, when they began a collaboration in New York that led to a joint Nobel Prize.

THE SECOND ROCKEFELLER FELLOWSHIP

In 1936, Max was already deep into biology. He joined Timoféeff and Muller on a trip to Copenhagen to attend a conference on radiation and mutation. Max, who was very enthusiastic about the mutation research, worked hard with Timoféeff to develop a better understanding of the underlying mechanisms. At the same time, he was facing difficulties within the German university system, after failing twice to meet the Nazi's political standards.

In this situation, Max was visited in 1936 by Dr. H. M. Miller from the Paris office of the Rockefeller Foundation, who indicated he was just checking up on former Rockefeller Fellows. Miller had visited Max before, in 1933, and both had agreed that efforts were necessary to fuse genetics and physics. The Rockefeller Foundation had concentrated on funding such cooperative research ever since Warren Weaver became director of its Natural Science Division in 1933. Among others, they supported the conference on radiation and mutation that Max attended with Timoféeff. Niels Bohr had proposed this symposium in 1935; it took place as the first "Physico-Biological Conference" in Copenhagen in 1936.

The Foundation eventually developed guidelines to support special developments in mathematics, physics, and chemistry that would contribute to biology; in 1938, Warren Weaver used the term "Molecular Biology" to describe this program. In addition, the Rockefeller Foundation had initiated a program for refugee scholars, that began in 1933 and would continue until 1945. The intent was to assist all scholars who had been displaced for political and racial reasons by the advent of Hitler. The Foundation established a Special Research Aid Fund for Deposed Scholars, and expended more than a million dollars for this purpose, supporting over three hundred scholars. The Rockefeller Foundation, sensing the war's imminence in Europe, advised the scientists they were interested in to choose between a perilous life in Europe and productive work elsewhere. Though not all recipients chose to come to the United States, the Foundation intended with this plan to invigorate American universities while also helping to protect European scientists.

When Miller visited Max in Berlin, the Rockefeller officer came to check whether Max was willing to leave Germany. At that time,

Max was studying the theory of natural selection developed by R. A. Fisher, J. B. S. Haldane, and Sewall Wright. Immediately, Dr. Miller suggested that Max apply for a grant from the Rockefeller Foundation to go to London to work with Haldane, Fisher, and C. Darlington. Max in turn wrote a letter to Haldane, asking if he would like him to come; as it turned out, Haldane was scheduled to be away from London for more than a year. At that point, Max came up with a new idea. On December 17, 1936, he informed the Rockefeller Foundation that he preferred another location to study genetics:

> However I have been wondering anyhow whether this plan which you suggested so unexpectedly, would be the most suitable thing to do. I thought that if I was to pursue work on mutations further, a stay in the United States where all the main work on mutations is being done would be more profitable. In particular, it would be of great help to me, if I could stay for some time in *Pasadena* in order to learn from T. H. Morgan and his co-workers, then in *Chicago*, to work and discuss Natural Selection with *Sewall Wright*, whose work I value very highly, and finally in *Cold Spring Harbour*, to discuss with Demerec his work on mutable genes, which is most interesting to Timofeeff and myself.

Max wrote this letter not to apply for a fellowship but simply to inquire as to Miller's opinion. The assistant director of the Rockefeller Foundation replied in early January 1937 that he very much favored these new plans, recommending that Max get busy in four different directions. First, the Foundation insisted on security; Max had to contact Lise Meitner to tell her to be prepared "to give the usual guarantees of a permanent post" to which Max could return. Second, he should prepare for a physical examination. Third, he had to get in touch with the scientists he wished to cooperate with. And fourth and most important, he had to obtain an exit visa so he could leave Germany.

On February 8, 1937, Max wrote to Thomas H. Morgan in Pasadena, to M. Demerec, Director of the Cold Spring Harbor Laboratory, and to the Minister of Education in Berlin. In the letter to Pasadena, he introduced himself and described his request:

> I am by training a theoretical physicist, but stimulated by Professor N. Bohr, I have for the last five years tried to learn so much of genetics and

of biochemistry that I might be useful in discussions where a good knowledge of atomic theory is required. . . . My work and study, being purely theoretical, will not require any special equipment.

Lise Meitner and Otto Hahn agreed, the physical check-up raised no problems, and in early June of 1937 the Education Minister gave Max a permit to travel to the United States. On June 13, 1937, Max informed Dr. Miller: "Today I am happy to be able formally to accept the Rockefeller stipend offered me through Professor Meitner in your letter dated April 27th." The Rockefeller Foundation had awarded him a fellowship in "Theoretical Genetics," paying once again $150 a month, with an additional sum for laboratory fees. Max was covered for travel from Berlin to Pasadena, via Cold Spring Harbor and Woods Hole, and for his return to Berlin. The goal of the fellowship was "to enable [Max] to study the theory of mutations, with particular reference to their origin by physical agencies."

This Rockefeller fellowship delighted Max; he felt immensely fortunate for the opportunity to leave Germany. As it turned out, Max did not return to Germany again during the war years. In retrospect however, he never thought it to his credit that he quit Nazi Germany. Many unpleasant things have been said about those who could have left Germany and didn't, the most outstanding case being Werner Heisenberg. Max never agreed with these derogatory judgments. In his view, many who stayed had imagined the Nazis would last only a short while; they felt it was important to see to it that some good people remained.

Max particularly admired the Bonhoeffers, who all stayed on in Germany, some becoming active in the Resistance. Max had placed science as his first priority; the German situation forced political compromises and distractions that he found distasteful. His family, however, remained to oppose the Nazis from both the liberal and the Communist side. Two commemorative stamps exist in memory of Delbrück relatives. One, issued by the East German government, honored Arvid Harnack and his American wife, who were both executed by the Nazis in 1943; a second, issued by the West German government, commemorated the theologian of the Resistance movement, Dietrich Bonhoeffer. He and his brother Klaus (Max's brother-in-law) were executed by the Nazis in 1945 for their resistance. Max's brother Justus had been imprisoned by the Nazis. He escaped as the Russians entered Berlin; later he was taken to a Russian camp, where he died of diphtheria.

GOING EAST

While Max left Nazi Germany in 1937 for the West, Timoféeff at first continued his experiments in Berlin. When he returned to the Soviet Union after the war, he was accused of having collaborated with the Nazis. As a consequence, Timoféeff was cut off from further genetics research. In addition, he fell on hard times as the teachings of T. D. Lysenko were declared official truth in the Soviet Union. Lysenko denied the entire gene concept, thus preventing any Soviet progress in the science of genetics until early 1966.

When the Lysenko debate came to an end, Timoféeff's situation improved at first, turning worse again when one of his assistants escaped to the West. Timoféeff was later awarded the American Kimber Medal for his scientific achievements, but he was not allowed to travel to the United States to receive the award. For the remainder of his days, Timoféeff stayed in the Soviet Union, leading an isolated and restricted life. He died there in 1981 (the same year as Max).

A LIVING MOLECULE

For Max, the thing most on his mind when he received the Rockefeller fellowship was that he could now seriously meet Bohr's challenge to find the paradox underlying biology. "A tantalizing discovery" (D 104) seemed to point the way. In 1935, the American biologist Wendell Stanley succeeded in crystallizing a virus in the same way that cellular molecules had been crystallized. The virus he was working with normally grew and multiplied in tobacco plants (Stanley, 1935). In other words, Stanley had found that a virus is a molecule, or as Max put it, "a living molecule" (D 104). This challenged the possibility of complementarity, since if a virus was indeed a living molecule, life was reduced to chemistry. H. J. Muller concluded from Stanley's work that viruses and genes had to be identical (Muller, 1936). He argued that viruses would offer the long-awaited simple system that would allow one to study replication.

During a farewell visit to Copenhagen, Max discussed this consequence of Stanley's discovery with Bohr. Max was certainly open to Muller's argument, since Bohr had always stressed the fact that physicists were fortunate to have had the simple hydrogen atom when they began to develop quantum mechanics. The virus could perhaps become the hydrogen atom of biology. A few weeks before his departure from Europe, Max formulated the potential lesson that virus research offered. He summarized his views of the riddle of life, sending his preliminary write-up to Bohr in August 1937. The German text was translated by Dean Fraser more than thirty years later, and was included in Max's Nobel address (D 76):

PRELIMINARY WRITE-UP ON THE TOPIC "RIDDLE OF LIFE"
(Berlin, August 1937)

We inquire into the relevance of the recent results of virus research for a general assessment of the phenomena peculiar to life.

These recent results all agree in showing a remarkable uniformity in the behavior of individuals belonging to one species of virus in preparations employing physical or chemical treatments mild enough not to impair infective specificity. Such a collection of individuals migrates with uniform velocity in the electrophoresis apparatus. It crystallizes uniformly from solutions such that the specific infectivity is not altered by recrystallization, not even under conditions of extremely fractionated recrystallization. Elementary analysis gives reproducible results, such as might be expected for proteins, with perhaps the peculiarity that the phosphorus and sulfur contents appear to be abnormally small.

These results force us to the view that the viruses are things whose atomic constitution is as well defined as that of the large molecules of organic chemistry. True, with these latter we also cannot speak of unique spatial configurations, since most of the chemical bonds involve free rotation around the bond. We cannot even decide unambiguously which atoms do or do not belong to the molecule, since the degree of hydration and of dissociation depends not only on external conditions, but even when these are fixed, fluctuates statistically from molecule to molecule. Nevertheless, there can be no doubt that such large molecules constitute a legitimate generalization of the standard concept of the chemical molecule. The similarity between virus and molecule is particularly apparent from the fact that virus crystals can be stored indefinitely without losing either their physico-chemical or infectious properties.

Therefore we will view viruses as molecules.

If we now turn to that property of a virus which defines it as a living organism, namely, its ability to multiply within living plants, then we will ask ourselves first whether this accomplishment is that of the host, as a living organism, or whether the host is merely the provider and protector of the virus, offering it suitable nutrients under suitable physical and chemical conditions. In other words, we are asking whether we should view the injection of a virus as a stimulus which modifies the metabolism of the host in such a way as to produce the foreign virus protein instead of its own normal protein, or whether we should view the replication as an essentially autonomous accomplishment of the virus and the host as a nutrient medium which might be replaced by a suitably offered synthetic medium.

Now it appears to me, that upon close analysis the first view can be completely excluded. If we consider that the replication of the virus requires the accurate synthesis of an enormously complicated molecule which is unknown to the host, yet though not as to general type, in all the details of its pattern and therefore of the synthetic steps involved, and if we consider further what extraordinary production an organism puts on to perform in an orderly way the most minute oxidation or synthesis in all those cases that do not involve the copying of a particular pattern—setting aside serology, which is a thing by itself—then it seems impossible to assume that the enzyme system for the host could be modified in such a far-reaching way by the injection of a virus. There can be no doubt that the replication of a virus must take place with the most direct participation of the original pattern and even without the participation of any enzymes specifically produced for this purpose.

Therefore we will look on virus replication as an autonomous accomplishment of the virus, for the general discussion of which we can ignore the host.

We next ask whether we should view virus replication as a particularly pure case of replication or whether it is, from the point of view of genetics, a complex phenomenon. Here we must first point out that with higher animals and plants which reproduce bisexually replication is certainly a very complex phenomenon. This has been shown in a thousand details by genetics, based on Mendel's laws and on modern cytology, and must be so, in order to arrive at any kind of order for the infinitely varied details of inheritance. Specifically, the close cytological analysis of the details of meiosis (reduction division) has shown that it is a specialization of the simpler mitotic division. It can easily be shown that the teleological point of this specialization lies in the possibility of trying out new hereditary factors in ever-new combinations with genes already present, and thus to increase enormously the diversity of the genotypes present at any one time, in spite of low mutation rates.

However, even the simpler mitotic cell division cannot be viewed as a pure case. If we look first at somatic divisions of high animals and plants, then we find here that an originally simple process has been modified in the most various ways to adapt it to diverse purposes of form and function, such that one cannot speak of an undifferentiated replication. The ability to differentiate is certainly a highly important step in the transition from the protists to the multicellular organisms, but it can probably be related in a natural way to the general property of protists that they can adapt themselves to their environment and change phenotypically without changing genotypically. This phenotypic variability implies that with simple algae like Chlorella we can speak of simple replication only so long as the physical conditions are kept constant. If they are not kept constant, then, strictly speaking, we can only talk of a replication of the genomes which are embedded in a more or less well-nourished, more or less mistreated, specific protoplasma, and which, in extreme cases, may even replicate without cell division.

There can be no doubt, further, that the replication of the genome in its turn is a highly complex affair, susceptible to perturbation in its details without impairing the replication of pieces of chromosomes or of genes. Certainly the crucial element in cell replication lies in the coordination of the replication of a whole set of genes with the division of the cell. With equal certainty this coordination is not a primitive phenomenon. Rather it requires that particular modification of a simple replication system which accomplishes constancy of supply of its own nutrient. By this modification it initiates the chain of development which until now has been subsumed under the title "life."

In view of what has been said, we want to look upon the replication of viruses as a particular form of a primitive replication of genes, the segregation of which from the nourishment supplied by the host should in principle be possible. In this sense, one should view replication not as complementary to atomic physics but as a particular trick of organic chemistry.

Such a view would mean a great simplification of the question of the origin of the many highly complicated and specific molecules found in every organism in varying quantities and indispensable for carrying out its most elementary metabolism. One would assume that these, too, can replicate autonomously and that their replication is tied only loosely to the replication of the cell. It is clear that such a view in connection with the usual arguments of the theory of natural selection would let us understand the enormous variety and complexity of these molecules, which from a purely chemical point of view appears so exaggerated.

Thus, before leaving Germany Max had begun thinking about the problem of reproduction in relation to physics. The fundamental property of an atom is its stability; the principal property of a virus is its multiplication. Max set out to find a system that could lead him to the bottom of this phenomenon.

The "Riddle of Life" stayed in Europe, while Max set off for the United States. Before he set sail for the New World, he wrote to Bohr that he was trying to follow advice given by Theodor Fontane. This German poet had described the painter Adolf Menzel in a short poem. After long hours searching for the poem in his father's library (Max first thought it was by Goethe), he finally found it and copied it for Bohr:

Gaben, wer hatte sie nicht
Talente, Spielzeug fur Kinder,
Erst der Ernst macht den Mann,
Erst der Fleiss das Genie.

(Gifts, who would not have them.
Talents, toys for children,
Only Seriousness characterizes a man,
Only diligence the genius.)

Of course, Max was ever diligent, but he was certainly not somber or staid in disposition. In that sense, seriousness did not characterize Max. One of the special qualities about him that attracted his friends was his sense of fun. He could throw himself into whatever he was doing; even science was pleasure. All his life, Max maintained the capability of enjoying science as if it were a game.

When he left Europe, Max was in a good mood. To him, this was not the end of one effort, it was rather the beginning of a new scientific adventure. Max never hesitated to begin something new. He felt much more at ease in opening a new field than in refining an advanced topic. Going to the United States gave him another opportunity to select a promising direction of scientific investigation. Max always enjoyed such a challenge.

He took with him something to keep him in touch with European culture: Jeanne Mammen's paintings. Eventually these would be exhibited in California, and become the "crowning glory" of Max's American home.

Molecular Biology in the New World

"All of a sudden, the whole world begins to take an interest in my viruses." (A photo created for a Nobel Prize reception at Caltech depicting Max and Salvadore Luria in ashcans, like Nagg and Nell of Beckett's Endgame. *Courtesy Seymour Benzer)*

The Atom of Biology

ARRIVING IN AMERICA

A handsome man, dressed in an elegant gown with beautiful patterns of trees and faces. He has lost one leg, which stands next to him. From his back, huge colorful wings emerge. He looks up, ready to fly.

This drawing, from an Ethiopian prayer book, dates from the early eighteenth century. Two hundred years later, it was on display in the Spencer Collection at the New York Public Library. Max purchased a postcard reproduction, carrying it around with him for a year until he finally mailed it to Jeanne Mammen, with the message that he was happy in the New World. He recommended the portrait to her as worth close inspection. Perhaps Max felt attracted to the man in the prayer book from Abyssinia since he too was anticipating going beyond the limits of knowledge in an emerging science. With his Rockefeller fellowship, he was going to study the theory of mutation, especially with regard to the influence of physical agencies.

In September 1937, Max had arrived in New York in a marvelous mood. The first thing he did was to visit the Rockefeller Foundation offices on the fifty-second floor of the RCA Building, where he found a wonderful view of midtown New York. When Victor Weisskopf later came to the city, Max picked him up at the harbor as he had five years earlier at the Copenhagen railway station. Max again greeted Victor enthusiastically, full of praise for America—especially for its apple pie and its street-numbering system.

After this splendid beginning, though, depression soon set in. Max left New York to visit genetics laboratories, intending to search for the ideal system with which to meet Bohr's challenge. During his farewell visit in Copenhagen, he and Bohr had discussed the potential of virus research in light of the recent crystallization of the tobacco mosaic virus (Stanley, 1935). Max had promised Bohr to spend the remainder of 1937 trying "to gain sufficient knowledge" *(Gelehrsamkeit)* so that he could improve on the preliminary write-up of the problem.

Max first spent about a month at the Cold Spring Harbor Laboratory, where he met with the director, M. Demerec. Cold Spring Harbor is a small village on Long Island; here the Carnegie Institution of Washington ran a Genetics Department. Max found Cold Spring Harbor irritating. The few people who were around during the off season were all especially quiet. As Max remembered, "nobody really talked to each other, so after closing hours there was just nothing you could do except go for a walk in the woods" (I). He was so disappointed by Cold Spring Harbor that in later years it took some effort to get him to return. However, the 1937 visit was definitely fruitful scientifically. In discussions with Demerec, Max learned of a new field, *Drosophila* cytogenetics. He marveled at the salivary gland chromosomes with their wonderful banding; Demerec even put Max to work in the lab dissecting *Drosophila* larvae, fishing out the salivary glands, and squashing and staining them. Quite probably, this introduction in Demerec's laboratory constituted the first experiments Max ever performed. Up to that point he had worked solely as a theoretician.

Although Max never continued experimentally with *Drosophila* genetics, he thoroughly enjoyed successfully repeating a standard procedure, and later required that all visitors to his laboratories conduct such experimenting to acquire experience on their own. By this, he did not just mean to introduce them to a particular experimental technique. For Max, experiments were fun, and he wanted to share this enjoyment.

He was happy to leave Cold Spring Harbor in November of 1937 for Rochester, New York, where he visited the German fly geneticist Curt Stern whom he had known in Berlin. Here, and later at Johns Hopkins University in Baltimore, Max learned how *Drosophila* chromosomes could be used to understand hereditary mechanisms. Not optimistic about the potential of this approach, however, Max found the methods complicated and the data unsuitable for rigorous analysis. This line of research did not offer him an immediate opportunity to apply his physical approach. Max's goal was to explain what a gene *does.* Having seen that physics had no trouble explaining what a gene *is,* he wanted to explore if physics could also prove useful in explaining the processes a gene has to undergo. His visits to the American laboratories constituted a search for the ideal system for studying replication—the basic process in living systems. Max had been optimistic about finding such a system in Princeton, on a visit with W. Stanley, whose work on viruses had aroused his interest.

The visit proved quite disappointing. In Princeton, Max experienced for the first time the contrast between a simple research idea and the reality of complex laboratory methods. The basic difficulty lay in the fact that in 1937, no easy way existed to analyze plant or animal viruses quantitatively. In order to study the reproduction of tobacco mosaic virus, for instance, one had to introduce the virus into a cell of the tobacco plant (Fig. 15). This had to be done by breaking off a cell through which the particle might enter, thereby creating an injury small enough to seal up afterwards. The method created virus assays with extremely low efficiency and provided insufficient data for a mathematical analysis. It would take another few months before Max eventually was to find a method he was satisfied with; the Princeton visit had not helped him.

15. *A leaf of Nicotiana glutinosa that has been inoculated with two forms of tobacco mosaic virus. Reprinted from The Natural History of Viruses, by Sir Christopher Andrewes. New York: W. W. Norton & Co., 1967.*

GOING WEST

After the visit with Stanley, Max set off for the West via Columbia, Missouri. Here he stopped to meet with Louis Stadler and Barbara McClintock, who were working on *Drosophila* and maize.

Louis Stadler, a "sort of counterpart" (I) to Timoféeff, had applied ultraviolet light to corn pollen, thereby discovering its mutagenic activity (Stadler, 1939). In discussing the details of these experiments, Max and Stadler got along very well. Such congeniality, however, did not arise with McClintock, the expert in corn genetics, who had worked out the relation between chromosomes and linkage groups in maize, showing that in corn, as in *Drosophila*, genes are located on these visible structures (McClintock, 1930; Creighton and McClintock, 1931).

McClintock's approach to science differed so much from Max's that even forty-five years later, she remembered there was no agreement between them. To her, the physicist aimed to take the organism to pieces and manipulate the parts. McClintock's goal was to gain an "inner knowledge" of what the organism can do under a variety of conditions, so that "it cannot fool you and you understand its answers." With that approach, she later discovered what she called the "controlling elements," namely, movable genes (McClintock, 1936). Her success came through observing the organism and noting that "one cell gained what another cell lost" (Fox-Keller, 1983). In 1983, McClintock was honored with a Nobel Prize for her work.

In a conversation in August 1982 Barbara McClintock still argued that Max never acquired that necessary inner knowledge of the organism he worked with, and that thus, "he didn't understand genetics, for he didn't understand his organism; it would fool him." Her criticism of Max's approach to science, an approach which originated in a realm of theory rather than observation, remains firm to this day.

ARRIVING IN PASADENA

After these visits to the Middle West, Max continued to Pasadena by train, arriving on the evening of October 15, 1937. At the station, he was met by G. H. M. Gottschewski, a German *Drosophila* geneticist, who first took Max for a beer and then dropped him at the

Athenaeum, Caltech's guest house and faculty club. Gottschewski quite upset Max by informing him that T. H. Morgan had thought it crazy for a theoretical physicist to come to Caltech to do biology. Having traveled 8,000 miles to reach Pasadena, Max found this comment unsettling. When he strolled in the vicinity of Caltech later that day, his despondency caused him to become utterly confused about directions: he couldn't tell north from south. Even decades later, the same Max who could find his way in a barren desert could not keep directions straight in Pasadena.

But when Max visited Morgan on his second day in Pasadena, he discovered that Gottschewski had misled him. Morgan was very cordial, recommending that Max work with A. H. Sturtevant, who in turn suggested that it would be interesting to try to clear up some confusing results on linkage in the fourth chromosome in *Drosophila*. Max was given some reprints to read. He couldn't understand them, for by that time the *Drosophila* terminology had become "so specialized and esoteric that it would have taken . . . weeks" (I) for a newcomer to master it. This was about twenty years after Sturtevant and Calvin Bridges had been able to provide direct cytological proof for the location of genes on chromosomes (Sturtevant, 1913; Bridges, 1916). Bridges's paper—"A Proof of the Chromosome Theory"—had formed the opening contribution in the new journal *Genetics,* the first of its kind in the United States.

Max was given a room across from Calvin Bridges. Here he sat, "poring over these papers pretty disconsolately for some time"(I); he "didn't make much progress in reading these forbidding-looking papers; every genotype was about a mile long, terrible" (I). Max never became excited by *Drosophila* genetics, though he regularly consulted with Bridges. They became good friends and would remain so till Bridges's untimely death in 1938 at the age of forty-nine. What fascinated Max was Bridges's "hippie" way of life. Everything Bridges did was utterly unpretentious and low key. This was a completely new experience for Max, who had never met such an easygoing person in Europe. Yet Calvin Bridges was an outstanding scientist, a "very wonderful *Drosophila* geneticist" (I) in Max's words.

The two men regularly lunched together. They walked to the corner of Lake Avenue and California Boulevard, a block west of Caltech. There they bought peanuts for ten cents and milk for five cents. They then strolled back to a bus stop with a bench, sat and ate their peanuts while discussing science and human nature.

GENETICS IN 1937

Max could not have hoped for a better mentor to introduce him to genetics; Calvin Bridges certainly was "among the leading geneticists of his time" (Morgan, 1940). He had been a forerunner in using the huge polytene chromosomes that had been found in the salivary glands of *Drosophila* larvae in order to construct a genetic map of these giant structures.

Bridges was part of the research group, later known as the Morgan School, which operated first at Columbia University and then from 1928 to 1945 at Caltech. During these years the group rapidly expanded knowledge of the physical basis of heredity. All their studies were conducted on one species of fruit fly—*Drosophila melanogaster*—since this organism was "so remarkably well adapted to the purpose of the human being who exploited it" (Dunn, 1965). In 1910, the group came together at Columbia where Bridges and Sturtevant joined T. H. Morgan as his students. This team stayed together and eventually moved to Caltech in 1928. Largely as a result of their efforts, genetics was close to its turning point when Max arrived at Caltech.

Genetics is a science of this century. Although Mendel had discovered the rules governing the hereditary processes in the 1860s, they didn't reach the biological community until their rediscovery at the beginning of the twentieth century. In 1906, the year Max was born, the term "genetics" was introduced by William Bateson in a book review; in 1909, Wilhelm Johannsen coined the names "gene," "genotype," and "phenotype." He thereby gave a name to the concept that had appeared in biological thinking before 1909—the view that the organization and the activities of living matter rest on a system of self-replicating living units.

The next decades saw rapid progress in genetics, chiefly through the study of *Drosophila* and maize. T. H. Morgan had initiated the work on *Drosophila* to address a fundamental paradox. In 1903, W. S. Sutton had argued that the hereditary mechanism involves chromosomes. Such a chromosomal basis was hard for Morgan to accept in 1910. He had begun his scientific career as an embryologist, and found it difficult to conceive how cells and organs could

differ from each other if their characteristics depended on chromosomes, and if these chromosomal structures in different types of
cells did not differ.

Eventually the work with *Drosophila* removed these grievous
objections, firmly establishing the chromosome as the basis of the
hereditary processes. This theory was widely accepted when, in the
late 1920s, H. J. Muller, another member of the Morgan School,
succeeded in inducing mutations in flies with X-rays. After this
breakthrough, which also stimulated Max's interest in the field, and
after it had been established that genes were arranged linearly on
chromosomes, Morgan came to see genes in physical terms. He
realized that in order to progress further, genetics needed help
from the physical sciences.

In 1932, Morgan outlined the future problems for geneticists
when he opened the International Genetics Congress, held in
Ithaca, New York:

> I have been challenged to state on this occasion what seems to be the
> most important problem for genetics in the immediate future. *First,* the
> physical and physiological process involved in the growth of genes and
> their duplication. The ability of the new genes to retain the property of
> duplication is the background of all genetic theory. *Second,* an interpre-
> tation in physical terms of the changes that take place during and after
> the conjugation of the chromosomes. *Third,* the relation of genes to
> characters. This is the explicit realization of the implicit power of the
> genes, and includes the physiological action of the gene on the rest of
> the cell. This is the gap in our knowledge. *Fourth,* the nature of our
> mutation process—perhaps I may say the chemico-physical changes in-
> volved when a gene changes to a new one. *Fifth,* the application of
> genetics to horticulture and animal husbandry.

In other words, a few years after Morgan had come to Caltech in
1928 to establish the biological sciences, he was planning what
today might be called molecular genetics. To his funding source,
the Rockefeller Foundation, he stressed his desire to integrate the
biological and physical sciences (1933); he pointed out that "the
success of this effort will depend in large part on the presence of
progressive and thoughtful men familiar with the most recent advances in physiology. The geneticists stand ready to cooperate"
(Kay, 1985; Allen, 1978).

So when Max arrived in Pasadena, he found some geneticists

ready to listen to his quantum mechanical model of the gene. He was soon asked by Sturtevant to lecture on the nature of the gene mutation, and in late 1937, he gave the first of many seminars at Caltech, in Room 101 in the Kerckhoff Building. This room is still used for seminars today. Next to it, in 103 Kerckhoff, Caltech has turned a former laboratory into the Delbrück Lounge. Max would have approved. He felt strongly that since scientists spend most of their wakeful existence at work, they should have a room for relaxation. Such a lounge would furthermore aid the free exchange of ideas. To Max, discussion was as important a part of doing science as running an experiment. Max would have helped make the lounge comfortable, bringing journals and books for coworkers to read, as he did in the small coffee room on his floor in Caltech's Church Building.

From the very beginning, Max made friends easily at Caltech. One of the earliest was Norman Horowitz, who in 1937 was working for his Ph.D. in embryology. Horowitz later joined the Caltech faculty. He still remembers the beginning of Max's first seminar. In spite of Morgan's support, Max's talk created considerable tension: here was a physicist who had hardly done an experiment in his life, coming to tell geneticists how to analyze a gene. When Max—this "strange creature," in Horowitz's description—opened his presentation by saying, "Let us imagine a cell as a homogeneous sphere," he was greeted by bursts of laughter. At the end of the seminar, Sturtevant thanked Max, assuring him that now they knew everything they wanted to know. No questions were asked following the talk; the gap between theory and experiment was too wide.

While reading up on *Drosophila* in 1937, Max found that classical genetics owed a great deal of its knowledge to the fruit fly. He did not agree, though, that the ongoing linkage studies in *Drosophila* could advance genetics understanding to a more mechanistic level. Max wanted to analyze the phenomenon of reproduction as a fundamental problem; he did not find *Drosophila* helpful.

PHAGE IN THE BASEMENT

Max later reminisced that if he had had no choice other than the fruit fly, he "would have been a failure," since there was no problem in sight that he could clear in a year's time, the duration of his fellowship. *Drosophila* seemed of no use to his dream of making use of physics in biology. Max was even ready to return to Germany and to physics. But Caltech's plant physiologist, Frits Went, encouraged him to continue looking for a simple system that might help to fulfill his vision. In order to discuss this further, Max accompanied Went on a camping trip to Arizona and New Mexico early in 1938.

When they returned, Max learned to his dismay that he had missed a seminar by Emory Ellis on bacterial viruses. Until then he had not known that anyone at Caltech was working with viruses. Max learned from this experience. From then on, he made a point of conversing widely in the institutions he was visiting or working in; he would ask researchers in great detail about their work, seeking to discover applications to his own interests and problems. No work was allowed to go unnoticed. Not only would Max discover what faculty members were working on, but he also made a point of talking to postdoctoral fellows and students. He even explained to each what his neighbors were doing. Max wanted to ensure that no others missed an opportunity, as he nearly did in 1938 when he stumbled upon the bacterial virus project at Caltech.

Max made up for missing the seminar by going down to the basement where Ellis was working to ask what had been accomplished in the virus studies. Ellis was analyzing bacterial viruses—creatures that prey on bacteria. They became known as bacteriophages, or, simply, phages. Max recalled being skeptical as he made his way downstairs since he had been told that Ellis was a biochemist who started "from zero knowledge concerned with anything about microbiology" (I). If anyone had told Max when he was on his way to meet phage that he was going there to stay, he would not have believed it.

Ellis greeted him cordially and showed him the equipment. Though everything was rather primitive, Max was impressed, since Ellis had started from scratch. And when he saw the first data Ellis had generated, he became very excited: "I was absolutely overwhelmed that there were such simple procedures with which you could visualize virus particles. I mean, you could put them on a plate with a lawn of bacteria, and the next morning every virus particle would have eaten a macroscopic one-millimeter hole in the lawn. You could hold up the plate and count the plaques. This seemed to me just beyond my wildest dreams of doing simple experiments on something like atoms in biology" (I).

Max did not hesitate to ask Ellis if he could join him in his phage work; he also went straight to Morgan to tell him that the bacterial viruses were so promising a system that he could spend the next ten years with them. Morgan did not put any obstacles in Max's path. In fact, Morgan's support later became crucial for the funding Max was to receive.

Max fell in love with phage for two reasons. First, one could generate suitable data for quantitative analysis overnight. One could ask a question one day, and have the answer the next day. Second, only rather simple equipment was required: an autoclave, a few dozen pipettes, petri dishes, and some agar. Max could immediately begin to carry out experiments himself, and soon move to experiments no one had ever done before.

Ellis had started to work with viruses after receiving a fellowship instituted by Dr. Seeley Mudd for cancer research. At the 1978 symposium celebrating the fiftieth anniversary of Caltech's Biology Division, Max described how Ellis had proceeded. After

reading about cancer, he found that there was a virus angle to it, and in reading about viruses he found out that there are bacterial viruses— viruses that attack bacteria; and he thought they might be more manageable for an experimental approach, especially by a one-man group. So he started reading about bacteriophages, and then he persuaded Dr. Morgan to let him get the equipment together: an autoclave which was about the size of a pressure cooker, a little sterilizing oven, 40 pipettes and about 40 petri plates. And then he went down to USC to his friend Carl Lindegren and obtained from him an organism that nobody in the Biology Division had heard of before, *E. coli*, and then he went down to the Los Angeles sewer department and procured 1 liter of sewage and filtered that and from that isolated a phage against the strain of *E. coli*. He taught himself how to plate this and how to get plaques.

Of course, *E. coli*, is now "the thing you hear about in grade school" (I). It has made gene technology enormously successful, producing drugs that can be purchased in pharmacies.

THE BACTERIOPHAGE IN 1937

Emory Ellis had started to work with *E. coli* for purely technical reasons (Ellis, 1966). His funding supported him to conduct research on cancer. In the 1930s, it had been shown that specific filterable viruses are causative agents of some diseases in plants, and of some cancerous growth in animals. They also could rupture some bacterial species in a process called lysis. The nature of these viruses was completely unknown, except for the virus that Stanley had successfully characterized in 1935. This virus caused the tobacco plant mosaic disease; it had been isolated in crystalline form and shown to contain two different types of molecules, known as nucleic acid and protein. When Max looked at the first plaque on a petri dish, it was not even clear whether viruses were particles or whether they represented some internal homogeneous change in the cells they attacked. All that was really known about viruses was that they were "subcellular entities capable of entering living cells and of reproducing only in such cells" (Stent and Calendar, 1978).

The term "virus" had been used in ancient Rome, where it meant poison of animal origin. During the Renaissance, this notion was refined to mean infectious agents responsible for contagious diseases. In the nineteenth century, some of these agents were identified as bacteria, but some remained invisible and elusive. They could even be passed through fine porcelain filters that retained all known bacteria. One of these "filterable viruses" was in 1915 shown not to be confined to reproducing in a higher cell; it could utilize lowly bacteria. This was discovered by F. W. Twort, whose publication remained unnoticed until, two years later, Felix d'Herelle announced the same observation: namely, a virus that grows on bacteria and kills them. He called this agent "bacteriophage" (from the Greek *phagein,* meaning to devour).

From the very beginning of his work on the bacterial viruses, D'Herelle considered phages particulate structures; this was still being debated in the 1930s. Max aligned himself with the particulate argument, immediately calling phages "the atoms of biology." This is obviously how they would appear from a physicist's point of view. When, in 1926, Albert Einstein was shown an experiment that demonstrated the lysis of bacteria by phage, he immediately recognized the particulate nature of phage. Felix d'Herelle described Einstein's response (D'Herelle, 1926):

> During my residence at the University of Leiden, in discussing this question with . . . Professor Einstein, he told me that, as a physicist, he would consider this experiment as demonstrating the discontinuity of the bacteriophage. I was very glad to see how this deservedly famous

mathematician evaluated my experimental demonstration, for I do not believe that there are a great many biological experiments whose nature satisfies a mathematician.

Such experiments not only satisfied Max, they overwhelmed him in Emory Ellis's laboratory in 1937.

D'Herelle invented the plaque-count method and applied dilution series for a quantitative study that allowed him to suggest a detailed picture of the life history of a bacterial virus (D'Herelle, 1926). D'Herelle's description of the phage life cycle is remarkably close to the more sophisticated picture revealed in the 1940s with the help of powerful tools unavailable to D'Herelle such as electron microscopy and tracer technology. Three steps were postulated: (1) The virus particle would attach itself to the susceptible bacterium; (2) it would enter the cell and multiply in it; and (3) it would disintegrate the cell, thereby releasing the progeny virus particles. This progeny would attach itself to other susceptible bacteria in the neighborhood and restart the cycle of infection.

In the early 1930s, only a few researchers accepted this notion. When Max arrived at Caltech, many scientists still regarded bacterial viruses as enzymes endogenous to the bacterium that were capable of catalyzing their own reproduction. It took quite a while for D'Herelle's description of the life cycle to be widely accepted.

Another contribution to phage research encouraged Max to embark on a study of these agents. By using various biophysical techniques, M. Schlesinger had succeeded in measuring the size of a phage particle, determining its linear dimension to be 0.1 μ. This was a more precise and smaller size than radiation studies had allowed Max to calculate for the size of his "gene." Thus in 1937, Max intepreted Schlesinger's phage to be a small gene. Today this is known to be wrong, since a typical phage may contain as many as several hundred genes. But in the late 1930s, both "gene" and "protein" were poorly defined molecular terms. It would take another seven years before the chemical nature of genes was identified. A better characterization of these macromolecules would only come in the early 1950s.

For a scientist in 1937, it was therefore impossible to distinguish the terms "phage," "protein," or "gene." Genes were assumed to be proteins, some of which were known to be autocatalytic enzymes. This was one way to interpret the multiplication of phage. To Max, it was of no importance to distinguish between phage, protein, and gene as long as he could study the key properties of a reproducing biological system that allowed an easy molecular quantification. In that sense, he was not interested in genes but in replication. The process was more important to him than the substance.

The holes that appeared in the bacterial lawn in phage's presence proved that phage was producing more of itself; it was self-replicating by consuming the bacterial lawn. Max believed that the best way to meet Bohr's challenge—to find a phenomenon that would *not* be explained without the help of complementarity—was to reveal the mechanism of replication in phage. As this had to be done carefully, Max immediately became critical. When Ellis showed him some data demonstrating that the growth of phage occurs in steps, he responded: "I don't believe a word of it" (Ellis, 1966, in PATOOMB). This is the first reported instance of Max's now famous reaction to new claims in science. He often responded this way, and he meant it, because he would only dispel this instinctive initial disbelief by his own analysis of the phenomenon. Max now prepared to do that with phage, designing an experiment to reveal the growth process of phage without any doubt.

ENTER M. DELBRÜCK

Students of biology read today that " 'Modern' phage research actually dates only from 1938, when M. Delbrück began to work with bacterial viruses" (Stent and Calendar, 1978). Max, in collaboration with Ellis, planned the so-called one-step-growth experiment (D 12). Its publication marks "the beginning of modern phage work" (Stent and Calendar, 1978). The procedure by which Ellis and Max analyzed the growth of bacteriophage remains the basic experiment for investigating phage multiplication (see Figs. 16a and 16b).

Their paper "cleared the mess," as Alfred Hershey expressed it more than forty years later; indeed, it caused Hershey to begin this type of biological research. A nice description is given by T. F. Anderson of the impact of Max's early phage research:

> During my three years (1937–1940) at the University of Wisconsin, I must have read many scientific papers . . . but today I can remember only three—one on growth of bacteriophage by Emory Ellis and Max Delbrück (1939) and two by Delbrück . . . [in 1940 (D 15, D 16)] on absorption and one-step growth. The experiments were beautifully designed and reported in an elegant style that was new to me. The three

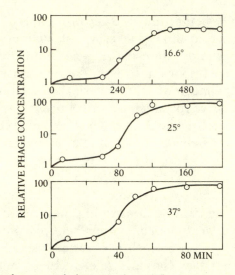

16a. One-step growth curves of phages at three different temperatures. This figure is taken from the 1939 paper by Delbrück and Ellis. The procedure is roughly as follows: bacteria and phages are mixed and incubated for a short time to allow adsorption to occur. The solution is diluted more than a thousandfold and incubated further. At certain times, samples are taken and put on a bacterial lawn for the plaque assay. The incubation is done at the indicated temperatures. The curves are S-shaped; the three component phases (latent period, rise period, and saturation period) are shown precisely below.

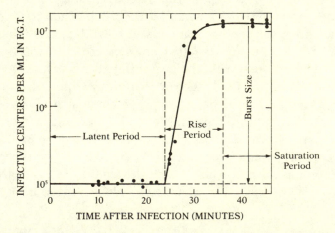

16b. A detailed one-step growth experiment allows one to distinguish three stages as defined in the diagram (Doermann, 1952). During the latent period, the phages remain within the bacterial cells. In the rise period, the phages emerge increasingly, until finally all infected bacteria are disrupted. In the saturation stage, one can determine the burst size. Reproduced from The Journal of General Physiology, 1952, vol. 35, p. 645, by copyright permission of the Rockefeller University Press.

papers carrying the Delbrück label formed a little green island of logic in the mud-flat of conflicting reports, groundless speculations, and heated but pointless polemics that surrounded the Twort-d'Herelle phenomenon. (Anderson, 1966)

What Max did was to refine the step-growth curves described by D'Herelle and devise experiments that allowed him to study the features of the phage life cycle separately.

In order to see how his background in the physical sciences allowed Max to move into phage research and advance it to productive experimentation, it is best to examine what is involved in a phage experiment. First, one incubates a solution containing a mixture of bacteria and phage in order to allow the phage to complete its life cycle as hypothesized by D'Herelle. A drop of this solution is then spread on a plate, along with sufficient bacteria to form a lawn. Each phage particle contained in this drop will destroy a bacterium on the plate, releasing phage progeny which will in turn attack and disintegrate neighboring bacteria (Fig. 17). Eventually all bacteria surrounding the one that was infected first will be disintegrated. After some time (overnight) this part of the bacterial

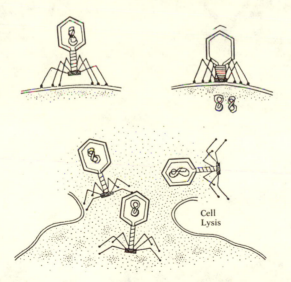

17. Summary of bacterial lysis of the bacteriophage T4, seen with receptors at the tips of its "legs." The receptors bind to the bacterial cell wall, viral genome is injected into the bacterium, and new phage are assembled there. The new bacteriophage are then released, as the host bacterial cell is lysed. Adapted from Biological Science, 4th edn., by William T. Keeton and James L. Gould. New York: W. W. Norton & Co., 1986.

lawn will look clear, with a visible hole in it. This hole is the plaque. One then counts the number of these plaques to obtain information about the number of viable phage particles (Fig. 18).

To put the experiments on safe ground, Max and Ellis needed a procedure to determine unambiguously the real number of viable phage particles in the suspension. Only this would allow full confidence in the plaque-count method. As a new technique, plaque counting needed credibility; they wanted to be sure it gave results comparable to those obtained by standard methods. With plaque counting grounded in this way, one would be able to proceed with measuring the principal parameters of the phage growth process. For example, one could then examine the number of phages that were released from a single bacterium.

The line of reasoning in this phage problem proceeded as follows: In a bacterial suspension, a single phage is enough to destroy all of the bacteria. The suspension that was turbid at the beginning will eventually clear. If one has made up a dilute phage solution

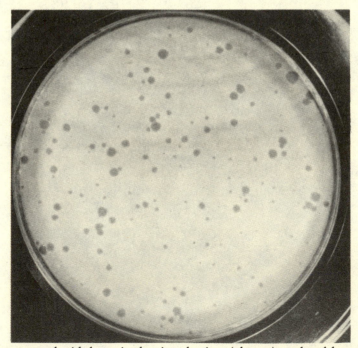

18. *Plate covered with bacteria showing clearings (plaques) produced by a phage particle. Reprinted by permission from The Natural History of Viruses, by Sir Christopher Andrewes. New York: W. W. Norton & Co., 1967. Photo by Dr. D. E. Bradley.*

and adds a small drop of it to several bacterial suspensions, one can determine the number of those drops that did not contain a phage particle by counting the test tubes that remain turbid. In those that clear, there had to be at least one phage. It is thus possible, by using a large number of drops, to determine the probability that a drop of a given size does not contain a phage particle. What one wanted to know was the concentration of phage in the solution, in other words, the average number of bacterial viruses in a drop. How could one derive this number from the number of drops without phage?

Such a problem was well known to physicists. A mathematical formula exists, called the Poisson distribution, to solve the problem. Since the application of the formula was crucial to the assay, Max searched for the original reference, as was his habit, jubilantly turning up the paper by Poisson (D 12; Poisson, 1837). This method allowed Max and Ellis to determine the efficiency of the plating of the phage particles, in other words the percentage of the phage in solution that gave rise to plaques. This in turn enabled them to characterize the way phages multiply in bacteria and to conceive the experiments that demonstrated the single-step growth of bacterial viruses.

The beginning of the phage success story was, in Max's mind, inseparably connected with an amusing story he related in 1978. It has to do with the difficulty of appreciating the virus assay. To see a plaque means actually to see nothing, that is, the absence of bacteria. That takes some serious looking. As Max narrated the incident, a few months after Max and Ellis had started their collaboration, "Ellis gave a seminar on phage [at Caltech] and he brought some petri plates along to show these plaques. Those were passed around and everybody said, 'Ah!' A few days later I met Mrs. Morgan and I asked her whether she was impressed with these [plaques]. She said, 'You know, the light was very poor. I couldn't see them.' It turned out that nobody had been able to see them. Everybody had taken it on faith that there were plaques there, which I thought was quite hilarious" (I).

Max's story goes on to illustrate his particularly characteristic response to this incident.

It reminded me of the story of the emperor's new clothes. I told Mrs. Morgan about it and she didn't know that story, so I made a special trip downtown to a secondhand bookstore [to see] whether I could dig up a copy of Andersen's Fairy Tales, and I did find a copy and bought it and took it home. It was an old copy of 1880 or something like that and the story was in there; then I looked at other stories, and the more I looked the more puzzled I became, because many of the stories didn't seem to be Andersen's fairy tales, but Grimm's fairy tales, which are an entirely different thing. Andersen's are invented ones, and Grimm's are folklore stories. Well, indeed, it turned out that the publisher of this book had simply stuffed his Andersen's fairy tales, which weren't enough but who was a popular name at the time, with some Grimm's fairy tales just for good measure. Those must have been the book publishing practices in America in the 1880s. Quite amusing. (I)

This incident not only points to Max's cultural breadth, his detailed knowledge of literature and sharpness of mind, but also demonstrates his love of education. If he felt something was important enough for an individual he dealt with to know about, he would personally provide the necessary help to get him or her started. In that sense, Max was an ideal teacher. He could always be relied upon to originate suggestions and to point the way; but he would never force anyone who demurred.

THE SECOND YEAR AT CALTECH

In August 1938, Max and Ellis submitted their paper on the growth of bacteriophage (D 12), which put the phage assay on safe ground but did not clear all the confusion around this phenomenon, as is evident from the opening sentence: "Certain large protein molecules (viruses) possess the property of multiplying within living organisms." The paper came just in time for Max to add it to his application for a renewal of the Rockefeller fellowship. The Foundation, committed to supporting theoretical biology, did not hesitate to grant another year. By the spring of 1938, Rockefeller staff had begun to wonder if they could find a position for Max in the States. The agent reported to the main office that "D[elbrück] is enthusiastic about his new field and I have the feeling that unless necessary, for economic reasons, he will probably not go back into pure physics. Our interest in developing the field of theoretical biology somewhat outweighs the technical difficulties in D's situation."

T. H. Morgan had praised Max's work to the Rockefeller office. Each time the Foundation inquired about Max, Morgan responded with enthusiasm. Morgan was convinced that the experiments with bacteriophage would yield the kind of data that a theoretical physicist needs for theoretical biological studies, and he recommended the renewal in April 1938: "We shall be more than glad to have him work with us for another year. It is not often that a physicist as competent as [Delbrück] is interested in applying his knowledge of physics to problems in biology. Moreover . . . he has made such a happy combination with Ellis that we can look forward to their turning out jointly a really significant piece of work."

After Max and Ellis "spent a marvellous year together" (I), to Max's regret, Ellis had to drop phage work and return to cancer research on transplantable tumors in mice. The fellowship supporting Ellis's work stipulated that his efforts be on cancer research. But Ellis continued to take an interest in what Max was doing after he was forced to discontinue the collaboration.

In the year that followed, Max's work on phage departed decisively from earlier attempts with bacterial viruses. Max approached the life cycle of phage as a fundamental problem of biology; he was not interested "in devising means to frustrate the growth of the viruses, [not] intent on applying such knowledge to the therapy of infectious diseases in plants, animals and men caused by viruses" (D 32). Max never left any doubt that "such motives, noble though they are, are ulterior to our cause" (D 32).

What he set himself to study was "the multiplication process proper." He wanted "to get to the bottom of what goes on when more virus particles are produced upon the introduction of one virus particle into a bacterial cell" (I). When Max described phage as the atom of biology, he probably was referring to the fact that physicists had only been able to explain the nature of matter by understanding features of atoms. He expected to find the solution to the riddle of life by investigating the reproduction of phage.

As a consequence, after Ellis's departure, Max went to work setting the stage for an analysis of phage as "a gadget of physics," to use one of his favorite expressions. He completed three papers, all appearing in 1940 (D 13, D 15, D 16). These publications brought him attention that swiftly led to collaborative efforts with others. Only in these three papers did Max work alone.

His starting point was the joint work with Ellis showing that a single phage attacks a bacterium; the single burst observed was interpreted as identical with the breakdown of the individual bacterium. "The main problem [was the] elucidation of the multiplication process" (D 12), which clearly occurred inside the bacterial cell. In order to be able to get data on replication, one had to know quantitatively how phage particles would be adsorbed by their bacterial hosts and what the role of the host's decomposition was in liberating phage progeny. In other words, in order to approach the study of replication as a physicist, Max had to calibrate his biological system. Then he would be ready to study "replication proper" (D 33).

During these studies, Max did not consider what was happening to the bacteria or how they were destroyed. For his purposes, he regarded the process as a black box, i.e., as a system, for which only the input and the output are analyzed and correlated (Olby, 1974). A single virus particle served as the input while the emergence of progeny was the output. Thus in his second Caltech year, Max was calibrating the black box *E coli.*

THE PRECURSOR IDEA

While Max was setting up phage to study the mechanism of replication, hoping it might enable him to answer whether or not virus multiplication was reducible to physical chemistry, he was also worrying about what other people were doing. It was still not generally agreed that phage growth was a multiplication process. Two biochemists, John H. Northrop and his associate at that time, A. P. Krueger, entertained the idea that when a bacterium is infected with phage, all the bacterial virus does is convert into phage a hypothetical pre-phage that the bacterium already contains within it. In other words, Northrop thought that "phage reproduction" was simply a chemical reaction rather than a basic biological process.

John Northrop was a famous American biochemist who had succeeded in crystallizing pepsin, a protein that can cleave other protein molecules (Northrop, 1939). By providing compelling evidence for the protein nature of his crystals, he provided the death blow to the picture of enzymes as small, organic catalysts loosely bound to carrier molecules. Northrop was awarded the 1946 Nobel Prize for Chemistry with James Summer, who in 1926 had crystallized the first protein ever (urease), and with Wendell Stanley. From the early twenties, Northrop had been publishing on viruses, too. In the mid-thirties, he isolated Staphylococcus phage and reported that the purified fractions contained nucleic acid (Northrop, 1939). Nevertheless, he posited an identical nature for enzymes and bacteriophages.

Northrop found that several enzymes, like pepsin and trypsin, are created as precursors such as pepsinogen and trypsinogen (Northrop, 1939). They are not enzymatically active, but become converted to the active enzyme when one active molecule cuts off a piece from the inactive form (zymogen activation). In other words, Northrop reasoned that what Max and Ellis had observed was not reproduction of phage and thus not replication. All they had observed was the activation of some precursors that become phage and activate other phages. This process would simulate reproduction.

Apart from the fact that Max would never have studied phage if the only thing he could analyze with it was the conversion of a precursor to a product, he also felt that

> Northrop, and especially his associate Krueger, basing their experiments on this preconceived notion of converting precursor into product, had done very bad experiments and published them, for several years, and had loused up the literature, and really confused the issue. So when we came along, the first thing we had to do was clean up this mess. And since these were people associated with the Rockefeller Institute . . . it took some hammering away. . . . I finally got their strain and did experiments on their strain, and showed that they were wrong, but I never got that published. I still have it in my file. For several years, I mean, my papers were largely directed at destroying Northrop and Krueger. (I)

THE FALL OF 1939

Max's second year in Pasadena came to an end in September of 1939. On September 14, he sent a report to the Rockefeller Foundation explaining how his "belief that the growth of phage was essentially the same process as the . . . reproduction of the gene" had led to successful experiments. He wanted to continue in this direction. This plan, however, was difficult to realize.

Originally, Max had been determined to return to Germany after two years. When he applied for the renewal of the first fellowship, he felt he could solve some of the obvious problems involved in phage study before the time came to head home; he even informed the Foundation then about his travel plans. He intended to return to Berlin by way of Japan. The Rockefeller people did not mind, as long as Max picked up the extra cost.

But Max was unprepared for the situation he faced in the fall of 1939. He had not clarified what was going on in phage reproduction, nor could he return to Germany. The outbreak of war made it virtually impossible for him to go back to Berlin, as he had agreed to do in the fall of 1937. Although Lise Meitner had given a guarantee for his job, the Nazis' aggressiveness against all Jews in Germany meant that even Meitner, an Austrian citizen, could no longer be protected. She had been dismissed in the summer of 1938, after almost thirty years of collaboration with Otto Hahn. She succeeded in leaving Germany to settle in Sweden.

Lise Meitner's fate had a strong impact on Max's plans. While at the end of 1938 he still favored returning to Germany, according to what he told the officials from the Rockefeller Foundation, by February 1939 he told Warren Weaver that he did not wish to go back to Europe. Weaver's diary paraphrases Max's reasoning: "His chief, Meitner, is gone, Europe is so unsettled and he wants to work in biology." In June 1939, Max asked the Rockefeller Foundation to assist him in finding a position in the States. T. H. Morgan assured the Foundation that Max "really understands biological problems" and advised them to help him find a job. Morgan wrote in September 1939:

He is not expatriated and not Jewish and, in theory at least, he could return to Germany. On the other hand I can say on my own responsibility that I think he is entirely out of sympathy with the present Nazi organization in Germany, and with his very liberal and democratic views he might very easily find himself in a tight place if he returned to Germany under the present conditions. He is not, then, in a technical sense a deposed scholar but he is undoubtedly a scholar under very distressing circumstances.

While he has, of course, been very circumspect in what he says about his political views, I know him well enough to be sure that he can never reconcile himself to the type of gov't at present under way in Germany. We have found him, during the two years he has been with us, quiet, gentlemanly, intensely wide awake to everything going on in the biological field to which he might apply his mathematical physics. He is one of the few men we have known who is a mathematician and to whom we can go with our biological problems and find that he has a real understanding of what we are trying to say. Under the circumstances I regret that we are so hard up that there is no possibility of finding a position for him in our department, which I would undoubtedly do were there any funds in sight.

He would consider very seriously any offer that should come to him and be prepared to undertake his responsibilities at once, although if it were possible without endangering his appointment he would like to have a couple of months to finish his work that he has been doing here.

Morgan urged the Foundation to help Max, "who is in a really critical situation." In October 1939, Morgan again turned to the Foundation's F. B. Hanson: "I asked [Max] what he was living on. He said his Fellowship had come to an end but that he had fifty dollars, and smiled."

Morgan very much wanted to help Max, whom he described as a "man with an unusual combination of mathematical physics and an interest in applying his knowledge to the borderland between biology and physics." But Morgan had just retired as a director of the Kerkhoff Laboratories at Caltech and his usual support was curtailed. This made it impossible for him to offer Max a position at Caltech. In addition, financial support for basic research was dwindling since the money was now needed for defense work for the war the country was about to enter.

THANKS TO ROCKEFELLER

Finding a job for Max was not easy. With training in physics rather than biology, he could more readily be hired to teach in a physics than in a biology department. Yet Max wanted to continue his work on phage. A solution appeared late in 1939. By that time, Max had begun living on money borrowed from friends. On December 21, 1939, he received a telegram from Francis G. Slack of Vanderbilt University in Nashville, Tennessee: "I am authorized to offer you a position in Physics Department at Vanderbilt at salary $2500 per year, starting January 2 or soon after as you can come." Max answered the telegram in typical fashion: he wrote a postcard saying he would be in Nashville on January 1, 1940. He went to Vanderbilt not as a professor but as an instructor of physics, and he remained in that position throughout the war years.

The contact between Vanderbilt University and Max had been made with the help of the Rockefeller Foundation. When the university inquired about the availability of a German physicist, since the language of physics before World War II was German, the Foundation informed Max. But the Rockefeller people did more than act as go-between. They made Max's position possible in the first place by approving an October 1939 request for aid by the chancellor of Vanderbilt, O. C. Carmichael. He had told the Foundation that Vanderbilt wanted to have Max on its staff, "but we do not have any funds for employing additional personnel in the dept. for 1939–40 session. If the Foundation could provide full salary for him for 1939–40, two-third salary for 1940–41, and one-half salary for 1941–42, the University would undertake to continue his service at its own expense after that time, provided, of course, that he should prove [acceptable] as a member of the staff."

ANOTHER TYPE OF COMPLEMENTARITY

In the summer of 1940, Max left Nashville to take care of his immigration status. Until then, he had been in the States as a visitor. To convert to immigrant status, he had to first proceed to Mexico and then return to Los Angeles. He decided to spend the necessary time in Mexico in Mexicali just south of the California border.

Before going there, Max met Linus Pauling on the Caltech campus. Pauling, too, was funded by Rockefeller; his chemical physics studies had the hope of leading to a greater understanding of the structure of substances important to biology. Max asked Pauling if he had read some recent papers by Pascual Jordan, one of the founders of quantum mechanics, who was a close friend of Max's from his Göttingen days. In his papers (Jordan, 1938, 1939), Jordan had claimed that quantum mechanics showed that identical macromolecules had a special quantum mechanical resonance attraction for each other, which might help to understand the mechanism of gene replication and the synapsis of like molecules on the strands of homologous chromosomes during meiosis.

Pascual Jordan, like Max, had been impressed by the complementarity argument, and in 1938 they exchanged a few letters about the relation of physics and biology. Max had written Jordan that he did not believe in these hypothetical pairing forces that would act between identical structures. He could not see any experimental evidence for them. He was not sure of his ground, though, because he felt insufficiently familiar with the application of quantum mechanics to complicated chemical systems.

When he told Pauling about Jordan's ideas, Pauling went over to the Physics Library with Max to examine the papers. After five minutes, Pauling called Jordan's idea "baloney." Max was impressed with the firmness of Pauling's opinion. When they met a few days later, after Max had been to Mexico and back, Pauling reminded Max of the Jordan hypothesis: "I have written a little note to *Science* about this; would you like to join me in publishing this?" Max mitigated the letter a little and signed it (D 17). Forty years later, Max explained his agreement by pointing out that he did not want to be "impolite" to Pauling. It seems more likely, though, that Max did not mind showing Jordan up. Max could be very impolite when he turned against what he considered bad science—and Jordan's work fell into this category, in Max's view.

When Bohr had advanced the complementarity argument, Max was not the only one who listened, though he was the only one to take it seriously enough to spend his life seeking its relevance. Jordan and Walter Elsasser also took up Bohr's challenge, attempting to apply the epistemological analysis to biology. Max always felt they confused the complementarity argument with the notion of vitalism. He complained in letters to Bohr and others that their "contributions . . . to biology have been *nil* (or negative, if you count the confusion generated in the minds of some of . . . [their] readers and listeners)." And he repeated the same argument as part of a review of Elsasser's 1971 National Science Foundation proposal ("Biological Individuality").

Ever since their joint note was published, Linus Pauling has claimed that Max and he stated as their firm belief that the principle of replication of the gene involved the synthesis of a complementary, rather than an identical, structure. This structural complementarity argument by Linus Pauling—he used the term "complementariness"—should not be confused with the deep epistemological argument by Bohr. The hypothesis advanced in the *Science* note did not strike any readers as having been very prophetic, though in the case of DNA replication, one has every reason to think that it is right.

ENTER LURIA

After this brief collaboration with Pauling, Max returned to his job in Nashville. In December 1940, he applied for his immigration

papers; he then proceeded to a meeting of the American Association for the Advancement of Sciences in Philadelphia. On December 28, he met Salvadore Luria there. Luria had been interested in
collaborating with Max ever since he had read the 1935 analysis of
the gene. He described their first encounter: "After a few hours of
conversation (and a dinner with W. Pauli and G. Placzek, during
which the talk was mostly in German, mostly about theoretical
physics, and mostly above my head) Delbrück and I adjourned to
New York for a 48-hour bout of experimentation in my laboratory
at the College of Physicians and Surgeons" (Luria, 1966).

The two men collaborated successfully in those forty-eight hours,
debating where and how to continue their joint efforts. On January
20, 1941, Max wrote Luria that he had been invited to attend the
symposium at Cold Spring Harbor and to spend the summer there.
He suggested to Luria that they work together there: "If that could
be arranged satisfactorily at C.S.H., I might overcome my antipathy
to the place." In the course of this collaboration, Max did indeed
come to fall in love with Cold Spring Harbor, and would eventually
spend his honeymoon on the grounds.

Insights from a Simple System

MUTUAL AFFECTION

In the early 1940s, Max's life changed—not only because he had moved from Pasadena to Nashville, but also because during that time he met the two most important people in his life. In 1940, while back in California, he was introduced at a party to Manny Bruce; in the same year, he met Salvadore Luria. In 1941, Max and Manny were married, and Max and Luria began their joint scientific work.

These events took place while Europe was beset by war; but Max was able to concentrate on science in America despite the situation back home. Thirty years later, in a letter to Jeanne Mammen, he remembered the Nashville experience: "Here I had the opportunity to develop my talents quietly while others were concerned with the war." With Luria's participation, molecular genetics was spawned in the secluded American South.

In his autobiography, Luria describes his first impression of Max:

> From the start Delbrück struck me as a dominant personality. Tall, and looking even taller because of his extreme thinness, moving and speaking sparingly and softly but in great precision, he conveyed the impression that whatever he said had been carefully thought out. His seriousness was occasionally broken by sparks of amusement, often produced by unexpected contrasts, especially by someone's pretentiousness. (Luria, 1984)

When Max met Luria, both were enamored of bacteriophage. It was, in Luria's words, "small enough to be like a gene and yet easy enough to work with" (Luria, 1984). Both men were very excited, hoping to produce in a short while all the data required for a complete analysis of their simple system. But their attack on the fundamental properties of life, though very successful later, took a little longer than anticipated. After the war, Max recalled his early enthusiasm and naive expectations in the Harvey Lecture delivered in the 1945–46 series (D 33).

The history of research on bacterial viruses or bacteriophages, now 30 years old, is fraught with controversy. The puzzling nature of these agents, living or inanimate, enzyme or virus, and the alluring possibilities of their useful application in medical practice have in times past attracted great scientists who spent their best efforts and keen imagination on devising ever new avenues of approach. This heroic age seems now a matter of the past, the passions have subsided and minor men with different interests are beginning to settle on these grounds. These minor men of our present age come from such far outlying fields as physics, genetics, biochemistry. They have a feeling that today many lines of biological research are converging on a central problem, the organization of the cell, a feeling similar to that which inspired the physicists of 50 years ago when the structure of matter and the constitution of the atoms became the focus on which all efforts converged. They feel that the field of bacterial viruses is a fine playground for serious children who ask ambitious questions. You might wonder how such naive outsiders get to know about the existence of bacterial viruses. Quite by accident, I assure you. Let me illustrate by reference to an imaginary theoretical physicist, who knew little about biology in general, and nothing about bacterial viruses in particular, and who accidentally was brought into contact with this field. Let us assume that this imaginary physicist was a student of Niels Bohr, a teacher deeply familiar with the fundamental problems of biology, through tradition as it were, he being the son of a distinguished physiologist, Christian Bohr. Suppose now that our imaginary physicist, the student of Niels Bohr, is shown an experiment in which a virus particle enters a bacterial cell and 20 minutes later the bacterial cell is lysed and 100 virus particles are liberated. He will say: "How come, one particle has become 100 particles of the same kind in 20 minutes? That is very interesting. Let us find out how it happens. How does the particle get into the bacterium? How does it multiply? Does it multiply like a bacterium, growing and dividing, or does it multiply by an entirely different mechanism? Does it have to be inside the bacterium to do this multiplying, or can we squash the bacterium and have the multiplication go on as before? Is this multiplying a trick of organic chemistry which the organic chemists have not yet discovered? Let us find out. This is so simple a phenomenon that the answers cannot be hard to find. In a few months we will know. All we have to do is to study how conditions will influence the multiplication. We will do a few experiments at different temperatures, in different media, with different viruses, and we will know. Perhaps we may have to break into the bacteria at intermediate stages between infection and lysis. Anyhow, the experiments only take a few hours each, so the whole problem can not take long to solve."

Perhaps you would like to see this childish young man after eight years, and ask him, just offhand, whether he has solved the riddle of life yet? This will embarrass him, as he has not got anywhere in solving the problem he set out to solve. But being quick to rationalize his failure, this is what he may answer, if he is pressed for an answer: "Well, I made a slight mistake. I could not do it in a few months. Perhaps it will take a few decades, and perhaps it will take the help of a few dozen other people. But listen to what I have found, perhaps you will be interested to join me."

A SERIOUS CHILD

Two terms meet the eye in this excerpt, which R. Olby characterized as Max's vintage piece (Olby, 1974): "childish" and "heroic." Science was Max's playground; he felt like a serious child, and struck others that way. That may be the best way to describe him. Science was an intellectual's game, to be played with enthusiasm. Max enjoyed thinking about science as a child enjoys a captivating game.

He also liked to play outside of science. If he wasn't working or reading, he would sit down for a game of chess, join friends in a game of cards, or play tennis. For Max, the outcome of the game didn't matter. Actually, he preferred to play against superior players, to improve his skill. He always wanted to do better, and he expected the same from others, in games as in science. One of Max's best chess stories shows him laughing at himself in a hopeless situation against a far superior player. He was playing against the mathematician Solomon Golomb, who could easily beat Max in spite of Max's sixty minutes per play to Golomb's single minute. When Max was utterly trapped, he looked at his friend and smiled: "Now I know how you do it. You think on my time." When friends inquired why Max could not beat Golomb in spite of his long deliberations, Max replied: "I think, but he knows."

Max enjoyed playing—not just winning—a game. Correspondingly, in science he enjoyed thinking about problems—not just solv-

ing them. He characterized himself as a childish young man entering the phage playground after the "heroic age" had come to an end.
"Heroic times" was the phrase used by Robert Oppenheimer to
characterize the early days of quantum mechanics in Berlin, Göttingen, and Copenhagen, the locations of Max's youth (Moore,
1985). It's typical that Max picked up the phrase. He always looked
at biology in analogy to physics, both historically and methodologi-
cally.

The latter is clear from the way he planned his experiments on
phage replication. In his plans, Max treated phage like a gadget.
However, the experiments he did on phage after he had calibrated
his system proved that the seemingly simple bacterial virus employed rather complicated mechanisms to achieve this elementary
result of replication. Here Max encountered for the first time what
he later termed "an ambiguity that pervades all of biophysics" (I).
What presents itself as a clear and intelligible behavior to the biologist is based on an intricate and interwoven development during
evolution. In Max's words, written about a decade after he moved
to Nashville, "any living cell carries with it the experience of a
billion years of experimentation by its ancestors. You cannot expect to explain so wise an old bird in a few simple words" (D 4).

MUTUAL EXCLUSION

When Max and Luria started their collaboration, both used the
same technique—the bacteriophage assay—and did the same experiments, although their goals differed. Max wanted to study reproduction, and that's what phage was doing in *E. coli;* Luria was
interested in the gene, and that's what phage appeared to be. However, in the early 1940s, no one knew for sure if these creatures had
genes at all. This first became clear in 1946, when it was discovered
that phage as well as bacteria have sex.

Independent of the possible conceptual difficulties, Max and
Luria shared one goal. Whatever phage was doing happened in the
black box *E. coli.* Max and Luria wanted to open this box first; in
order to do so, they conceived of the mixed-infections method.
They would add two different types of viruses to one bacterial
strain. One type was slower than the other; that is, it would take

longer to break open (lyse) the bacterium. Their first joint publication (D 20) describes the experiment:

> The growth of a bacterial virus (Bacteriophage), occurring only in the bacterial cell, may be said to proceed behind a closed door. The experimenter can follow the virus up to the moment it enters the cell, and again after liberation from the cell. There is, as yet, no way of telling what goes on within the cell, except by circumstantial evidence which covers the entry of the virus into the host, its time of stay, its exit, and, perhaps, the metabolism of the host cell.
>
> By the desire to gain more direct insight into the intracellular processes of virus growth, the present authors were led to try the simultaneous action of two different viruses upon the same host cell. There was a possibility that one virus might lyse the cell, while the other was still growing. Thus an intermediate state of virus growth would be revealed. This expectation did not materialize.

Max and Luria discovered that two bacterial viruses acting upon the same host interfere strongly with each other's multiplication. When a bacterium is simultaneously infected with two or more bacteriophages, only one type of virus will be liberated from any one cell. The discovery of interference dealt a first blow to Max's hope that replication in phage might be a simple phenomenon. The growth of bacteriophage clearly interacted in some complicated way with the bacterium. In spite of this, Max felt that the investigation of mixed infections was "a lucky choice of problem," since the discovery of mutual exclusion "put into striking relief the fundamental similarity of bacterial viruses with animal and plant viruses," as Max wrote to the Rockefeller Foundation in 1945. He could be sure now that he was indeed "dealing with an aspect of the relation between viruses and their host cells of universal validity" (D 31).

Max was so excited about this "novel phenomenon" that he introduced more meaningful terms for it. In his third publication on interference (D 31), produced without Luria, he suggested the name "mutual exclusion effect"; in 1947, in a German publication, he used the phrase "the principle of mutual exclusion." The terms "interference" and "mutual exclusion" are taken straight from physics. As indicated by the titles that he gave two of his most important lectures (D 41, D 76), Max considered himself a physicist spending his time in biology. The interference problem seemed to offer a possibility to make the life sciences a branch of physics. Max continued to study this phenomenon even though Luria lost interest in it rather early, since mutual exclusion did not help Luria in analyzing the nature of the gene. It was, though, an accessible part of the growth of phage; that was Max's interest.

Max kept at this problem for nearly ten years (D 44) during which time, in collaboration with Jean Weigle, he analyzed the mutual exclusion between an infecting phage and a carried phage. This phenomenon, though not yet fully explained, seems to have become a forgotten subject nowadays. Its complementary effect— recombination—is considered much more important. A few years after the exclusion effect was observed for the first time, certain so-called mixed infections were found not only to violate Max's principle, but even to give rise to newly recombined phage particles (Hershey, 1946). Molecular genetics was ready to explode.

THE DELBRÜCKS

After the first summer which Max spent with Luria in Cold Spring Harbor, he did not return to Nashville but instead went to California to get married. As he had told his friend Milton Bush, "I can no longer do without it." On August 2, 1941, Max married Mary Adeline Bruce, the daughter of mining engineer James Latimer Bruce and Leah Hills. Manny—as Max's wife is called by all who know her—doesn't remember much of their honeymoon, except that "he couldn't wait to get back to Cold Spring Harbor." Manny and Max had met the year before when Max had come to Pasadena to change his immigration status; both had been invited to a party at Arnold Bergstrasser's home. Also a German émigré, Bergstrasser taught at Scripps College in Clarement, where he had directed Manny's undergraduate thesis on Michelangelo.

Born in Butte, Montana, Manny grew up on Cyprus, where her father managed copper mines owned by the Mudd family of Los Angeles, and went to high school in Beirut, Lebanon. When she reached college age, the family moved to the Los Angeles area and Manny entered Scripps.

Max immediately distinguished himself at the party at Bergstrasser's home by reading some poems by Goethe to Manny, entertaining her with his interpretations. Manny was impressed by this interest in poetry; she felt that life might not be uninteresting with such a man. But she quickly discovered Max was not conventionally romantic in nature. His first present to her arrived on Valentine's Day, 1941, in a package from Nashville. Wondering what it might be, she carefully opened it to discover a clothesbrush. When she later asked him why he had selected such a present, Max answered that he had never seen such a cute clothesbrush before and had figured that she would like it.

Before their marriage, Manny found Max nervous and unsettled, since he had actually intended to return to Germany and only decided to emigrate later, after the war started. But coming from the strong, very organized family of Delbrücks, Max was a "family-conscious" man. Manny gave him just what he wanted, harmony and grounding in America. She sheltered him, allowing him to concentrate on science. Manny ran the house, looked after the car, did the income tax, and later on guided the children.

In contrast to Max, who never owned a television, read magazines or took much notice of the outside world, Manny became involved in local politics and tried to influence matters according to her liberal views. She is a good writer, and while they lived in the South, contributed to local newspapers. After they had been married for a year, she wrote what she called "A Silly-Dilly for Max," reporting on her studies of a new species: Homo Scientificus. This branch of the family of man is, as Manny notes, "easy and interesting to observe, but difficult and perplexing to understand." What struck her most was the way the male Scientificus approaches the female. He "does not . . . delight in flaunting elegantly before the female to catch her eye"; rather, he "brings her a little gift such as a bundle of bristles or a bright piece of cellophane, which she accepts tenderly, and the trick is done." The Homo Scientificus, according to this one-year study, "is one of the best playing animals known," and it is not difficult to satisfy him. He is content "so long as he can spend most of the day sitting in the sun and rummaging among his strange possessions."

In addition to her writing, Manny also worked part time at Vanderbilt University. Through her contacts, they made quite a few Southern friends. One such contact helped to relieve Max and

Manny from financial difficulty. In a 1978 interview, Manny recalled this episode from their life in Nashville:

> Socially, we consorted with a small group of intellectuals who were a fairly strong influence in our lives. The leader of this group was a Jewish businessman, Alfred Starr, a very bright, energetic person who owned all the Negro theatres in the South—fairly well-to-do. Alfred Starr felt that it wasn't right for Max to still be an instructor at Vanderbilt. The reason, I think, that Max was still an instructor was that he was German, and they didn't want to make him a professor. But Alfred Starr went to the provost, and said he wanted to pay the difference between Max's salary and that of a full professor. So he did that, and then we got $4,500 a year, which was as much as anybody got then.

Starr thus approximately doubled Max's income. Originally, when the Rockefeller Foundation supported the move to Vanderbilt, they had agreed to pick up part of Max's salary so he would earn "not less than $2500." This arrangement with the Foundation was supposed to terminate in 1942. But on May 1, 1942, Chancellor O. C. Carmichael of Vanderbilt called the offices of the Rockefeller Foundation, which at that time had a special department for what they called Molecular and Mathematical Biology. The chancellor informed the Foundation that the university could not pick up Max's salary *and* research expenses, because of the reduced income from student fees and also because Max's "research expenses are increasing, due to the success of his work." The Foundation approved action providing funds for the research expenses, while Vanderbilt assumed the entire salary.

This grant from the Rockefeller Foundation expired on June 30, 1945, about the time World War II came to an end in Europe. Whereas World War I had been a disastrous shock to Max, the second one went seemingly unnoticed. Luria doesn't remember any serious discussion between the two of the horrible events that were ruining their home countries.

It was clear, though, that the war created difficulties for Max. He was, after all, a German citizen living in a country fighting a war against his country. As Francis Slack, then chairman of the Physics Department, remembers, this led to some resentment by Nashville citizens. The scientists didn't care that much; they considered Max more as a European than a German. But the people in town were upset that he did not immediately apply for American citizenship.

One scientist who became a good friend of Max's was Milton Bush, a professor of pharmacology. They met for long tennis matches during the day and chess games in the evening. Milton did not recall any instance in which Max discussed the war. He was also certain that Max was completely unaware of the activities in nearby Oak Ridge.

Being in the South during the first half of the 1940s was very comfortable for Max. He was little hindered by the war, and there weren't enough aliens from enemy countries in the South to intern, as occurred for Germans and Japanese in California. Max did not even realize the hostilities might affect scientific activity. While he—a German—was working with Luria—an Italian—Max received a request from Japan to teach them some phage techniques. One year after the Japanese had bombed Pearl Harbor, Max inquired at Vanderbilt if he could invite a Japanese microbiologist to join his laboratory. This request was, not surprisingly, denied by the university.

During the early forties, when he taught physics in Nashville, Max was mentor to only one graduate student. In 1942, Ernest Jones wrote his master's thesis on the ultraviolet rotatory power of crystalline potassium bromide. Jones's memories of Max were not flattering; he recalled a "poor teacher of physics," who did not really feel at home with complicated measuring devices. Max, of course, was distracted from teaching by the rapid progress of his phage work.

A LOOK AT PHAGE

The early 1940s saw the introduction of a new and powerful technique in biology: the invention of the electron microscope, which allowed examination of phages. This closer look revealed not only molecular beauty but a new level of complexity. The electron microscope converted the objects of biological research from abstract symbols to concrete entities with morphological individuality.

Max entered the "picture-taking business" during the summer of 1942 when RCA installed an electron microscope for T. F. Anderson at the biologists' "summer camp," the Marine Biological Laboratory at Woods Hole. Anderson offered to collaborate with anyone wanting to use this instrument. He had begun joint work with Luria in the spring of 1942 and Max would follow later that summer (Fig. 19). When Luria managed to produce stock solutions that assayed as high as 10^{10} phages/ml, they were able to visualize the phages, which they found to be tadpole-shaped particles whose heads ranged from 600 to 800Å in diameter (Luria and Anderson, 1942) (Fig 20). They had no doubt that they had found phage particles.

Luria and Anderson showed their electron micrographs to J. J. Bronfenbrenner, then of Washington University in St. Louis, who "clapped the palm of his hand to his forehead and exclaimed, 'Mein Gott! They've got tails!' " (Anderson, 1966, in PATOOMB). Bronfen-

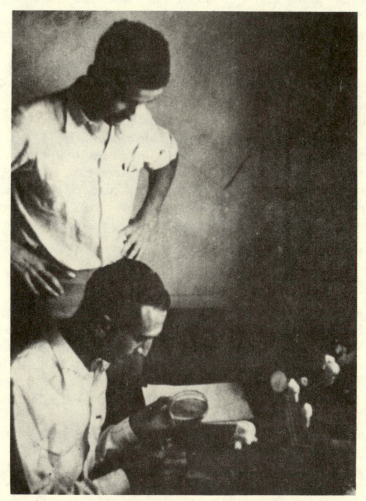

19. *Max and Luria in the lab at Cold Spring Harbor (1941). Reprinted by permission of Cold Spring Harbor Laboratory.*

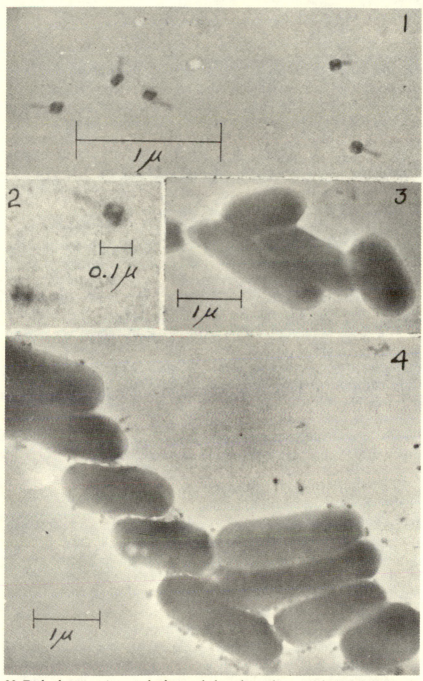

20. *Early electron micrograph photos of phage by Anderson and Luria, taken at the RCA research laboratories in late 1941 or early 1942. Plates 1 and 2 show the characteristic round head and thin tail of phage which gives a spermlike appearance. Plate 3 shows* E. coli *in distilled water, while Plate 4 shows* E. coli *in a suspension of phage for 10 minutes. Visible sperm-shaped phage particles have adsorbed on the* E. coli *cells. Photos from Proceedings of the National Academy of Sciences, 1942. Reprinted courtesy of Dr. Salvadore Luria.*

brenner was the first American to work with phage after it had been discovered in Europe. Having inferred incorrectly from his diffusion experiments that phages were much smaller particles than they turned out to be, he was absolutely astonished by some details of the structure.

The findings that Luria and Anderson reported in 1942 had already been made, it turns out, by H. Ruska in Germany (Ruska 1940, 1941), but the war had prevented any communication. Ruska was awarded the Nobel Prize for Physics in 1986.

When Max joined Luria and Anderson for a few weeks in the summer of 1942 at Woods Hole, he wanted to find out whether the electron microscope might allow a closer look at interference. In their joint experiments, they paid great attention to controlling two factors quantitatively: the concentrations of bacteria and phage, and the time during which they interact. In that way, they could be a little more precise as to the adsorption process. The electron microscope work confirmed some of Max's earlier conclusions by demonstrating the adsorption of the phages on the host cell and, at the predicted time, the disruption (lysis) of the host cell and the expected liberation of a hundred or so daughter particles from each cell. The pictures showed also that the phage never entered the cell; it remained outside during its life cycle. The growth of phage was much more complicated than Max had anticipated.

THE BIRTH OF BACTERIAL GENETICS

In February 1943, Anderson mailed his joint paper with Max and Luria to the *Journal of Bacteriology.* By that time, the European duo—Max and Luria—had just worked out the result that would make them famous. They had found a way to prove the spontaneous nature of bacterial mutations. The publication of Max and Luria's observations is considered a landmark in the history of biology. In the words of a well-known introductory text in molecular genetics (Stent and Calendar, 1978), "Just as the birth of genetics is considered to have taken place in 1865, upon the appearance of Mendel's paper, so the birth of bacterial genetics can be dated 1943, when S. E. Luria and M. Delbrück published a paper entitled 'Mutations of Bacteria from Virus Sensitivity to Virus Resistance' " [D 25].

The observation that led them to analyze the mechanism by

which bacterial variants appear had already been made in 1939. In his final report to the Rockefeller Foundation, Max described a new phenomenon. In the last paragraph written on September 14, 1939, he stated:

> In another series of experiments it was attempted to clear up the origin of the secondary growths which always appear a few hours after a susceptible culture of E. coli has been lysed. It was found that this secondary growth is a resistant strain which arises by mutation from the susceptible strain. The mutation rate is about 10–5 per generation. Reverse mutations have not been observed. In a mixture of resistant and susceptible bacteria the susceptible ones are at a slight selective advantage, so that they will outgrow the resistant fraction over long periods of time. This explains why a susceptible strain remains susceptible although there arise constantly resistant mutants.

This observation indicated very early to Max that phage replication could not be looked at as a simple isolated process. Phage was interacting with bacterial organisms; it exhibited complicated reaction kinetics and showed an amazing morphology. Max's strength was that he did not hesitate to embark upon a new problem. By following up on this observation of "secondary growth," even though it interfered with his ideal of simplicity, Max made a major contribution to bacterial genetics for which he is remembered today.

The appearance of resistant bacterial variants could easily be observed. A bacterial culture that is attacked by a bacterial virus turns clear after the bacteria are lysed. Upon further incubation, the solution becomes turbid again in a few hours due to the growth of the resistant bacterial variant ("secondary culture"). The real difficulty was to determine whether resistance was the result of a spontaneous mutation in the bacterial line, or whether resistance somehow arose because of the continued presence of the virus. The later hypothesis, which was called adaptation, might be explained by the appearance of new enzyme systems by non-genetic means whose formation was induced by components in the culture medium. Any observed change in a physiological characteristic could be explained by either a mutation or a so-called adaptation; it was imperative to distinguish sharply between these two possibilities.

In 1934, I. M. Lewis had hypothesized that a bacterial variation was caused by a mutation. Colon bacteria grown in glucose usually could not ferment lactose. Upon close inspection, Lewis observed that one individual in a million (10^6) could change, to be able to live on the disaccharide. Lewis concluded that a mutation had occurred, but he had no proof; he could not rule out the alternative explanation. This was finally accomplished by Max and Luria in 1943, four years after Max had made his first observation of secondary growth.

Two years after first noting the growth, Max mentioned it again in a note to Luria in April 1941: "I agree with you that the rise of the secondary culture would be an interesting and attackable problem for future collaboration." They drew up plans for a cooperative research project. After their meeting during the summer of 1941 in Cold Spring Harbor, Luria applied for a Guggenheim fellowship to continue the collaboration with Max at Vanderbilt. When this was granted, they continued to work on the problem in 1942.

Max described the history of their seminal publication (D 25) in a report to the Rockefeller Foundation on June 6, 1945:

> We agreed that he should work on the problem of the origin of bacterial variants which are resistant to the action of a given virus and which occur in practically all sensitive cultures. The question had never been settled whether these variants are produced in response to the action of added virus or whether they occur spontaneously, and are brought to the foreground through the selective action of the virus.
>
> This problem proved harder than we had expected. All attempts at devising critical experiments were unsuccessful during Dr. Luria's stay here at Vanderbilt during the fall term of 1942. However, Dr. Luria continued work on this problem at Indiana University, where he had accepted a position as instructor in bacteriology, and early in 1943 hit on a satisfactory method of deciding the issue. This method was worked out, the theory by me here at Vanderbilt, the experimental technique by Dr. Luria at Indiana, and the work was completed during another visit of Dr. Luria to Vanderbilt during the spring of 1943.

Luria has described how he hit upon the "satisfactory method": "The idea . . . came to me, in fact, while watching the fluctuating returns obtained by various colleagues of mine gambling on a slot-machine at the Bloomington Country Club, where faculty dances were then held one Saturday a month." (Luria, 1966, in PATOOMB)

Luria had started an experiment which was designed to find out whether the number of bacteria resistant to phage attack does rise in a growing *E. coli* culture. He had reasoned that if the resistance was induced by the presence of phage, then at all stages of growth the fraction of resistant bacteria should be the same. If the resistance was due to a spontaneous mutation, the new variant strain would steadily create new mutant lines, called clones. These clones would divide at the same rate as the sensitive bacteria; thus, the number of resistant variants would continuously rise with the age of the culture.

As he actually did the experiments, Luria found to his annoyance that the number of resistant *E. coli* cells was subject to great day-to-day fluctuations. When he watched the slot machine, it occurred to him that he was analyzing the wrong numbers. Instead of thinking about the number of resistants, he felt he should look at the fluctuations. Luria now moved the horse to its proper position before the cart. In a letter to Max on January 20, 1943, he recommended that they compare the fluctuations in two experiments:

I thought that a clean cut experiment would be to find out how the fluctuations in the number of [phage-]resistants depend on the culture from which they come. That is: If I plate with . . . [phage] ten samples of the *same* culture of [*E. coli* strain] B, I find numbers of resistants which fluctuate according to Poisson's law. If I plate 10 samples of 10 *different* cultures of [*E. coli* strain] B, all containing the same amount of B, I find much larger fluctuations. If the resistants were produced on the plate, after contact with . . . [phage], they should show the same fluctuations in both cases.

Max responded on one of his famous postcards, dated January 24, 1943:

You are right about the difference in fluctuations of resistants, when plating samples from one or from several cultures. In the latter case, the number of *clones* has a Poisson distribution. I think what this problem needs is a worked out and written down theory, and I have begun doing so.

On February 3, the manuscript with the theory arrived, and Luria began experiments analyzing the fluctuations of *E. coli* mutants resistant to the phage now called T1. Now things moved swiftly. In May, Max and Luria submitted their proof of spontaneous mutations of bacteria from virus sensitivity to virus resistance. It was published in the November 1943 issue of *Genetics,* which actually did not come out until early 1944.

With this publication, bacterial genetics begins. For though Max and Luria were concerned with the properties of bacteria, they did not publish their paper in a bacteriological source but in a genetics journal. In 1943, geneticists did not generally read papers on bacteria and bacteriologists did not think in terms of genetics. In addition, apart from providing adequate evidence that phage-resistant mutants originated by spontaneous mutations, the fluctuation analysis provided the means to determine mutation rates. In Luria's words, "the value of the fluctuation method is precisely that *it makes the fluctuations themselves the basis for the measurement of mutation rates"* (Luria, 1966, in PATOOMB). This can be done using Poisson's formula.

Max and Luria's paper provided statistical evidence for the existence of genes in bacteria and for their spontaneous mutation. These findings by themselves could not lead to the explosion in bacterial genetics, for the sex life of bacteria still needed to be discovered. Edward Tatum and Joshua Lederberg achieved this in 1946.

CRITICAL CHOICES

The importance of the 1943 fluctuation paper rests on two further grounds. First, in Gunther Stent's words, the resistance paper "did for bacterial genetics what Mendel had done for general genetics— namely showed for the first time what kind of experimental arrangement, what kind of data analysis, and, above all, what kind of sophistication was needed for obtaining meaningful and unambiguous results . . . their paper became the standard by which all later papers on bacterial genetics were to be measured." (Stent, 1981).

The second point may be even more important. The two mechanisms Max and Luria suggested to account for heritable changes in the properties of bacteria are somewhat analogous to the two major views of evolution, those of Darwin and Lamarck. The term "Lamarckism" refers to the belief in the inheritance of acquired characters. It assumes a mechanism by which changes in the environment can induce changes in a cell which are then passed on to succeeding generations. Darwinism is a theory of evolutionary change based on mutations that occur at random; nature acts on (selects from) the results of these spontaneous genetic changes.

Lamarckism was the view held by most of the biologists before Darwin published *On the Origin of Species* in 1859. Now, of course, scientists universally accept a view of evolution that is consonant with what Darwin proposed. By using the fluctuation test to establish that phage resistance results from spontaneous mutations, Max and Luria removed bacteriology from what Luria called "the last stronghold of Lamarckism."

Proponents of the adaptation theory maintained even into the 1950s that Max and Luria's results could be explained by different assumptions (Jackson and Hinshelwood, 1950). In order to do so, they posited a hypothetical "lag phase," during which no growth should take place. Bacteria were supposed to experience such a phase when introduced to a new environment. Such a lag phase complicates matters in an unnecessary way and therefore lends credence to the mutation hypothesis; it is difficult, however, to discriminate between the lag phase hypothesis and the mutation hypothesis where resistance to bacteriophage is concerned (Armitage, 1952).

COMPLEMENTARY POINTS OF VIEW

The fluctuation test gave microbial genetics one of its big boosts. It allowed the measurement of lower mutation rates, by several orders of magnitude, than had hitherto been possible. As a result, it attracted attention to the remarkable possibilities of bacterial genetics.

With such resounding implications being drawn from their paper, Luria felt as if he "was literally walking on clouds" (Luria, 1984). His interest in the nature of the gene now led him to embark on mutation research, and he eventually isolated and analyzed the first so-called host-range mutations of bacteriophage (Luria, 1945).

Max took a different path. He called their milestone in the history of genetics "a side issue which is threatening to displace the main issue, by virtue of its explosive content of possibilities for studying bacterial genetics" (D 33). The main issue was the replication of phage; the bacterial variants "had nothing to do with it." Max continued to analyze mutual exclusion in mixed infections. While he considered the fluctuation paper "interesting," he felt the "interference experiments were great fun" (I).

At first glance, it seems strange that Max, the founding father of molecular genetics, did not show any interest in exploiting the study of mutations. Such an attitude would surface again when he switched to studies of a unicellular fungus in the 1950s. To Max, who had studied them as far back as 1935, mutations were the handle to get hold of the molecular workings of the gene. And yet no one in 1943 could be sure that bacteria or their phage had genes at all. This question was not answered until 1946. Max did not seem to want to be a geneticist; he was more comfortable as a physicist investigating biological problems.

ENTER HERSHEY

While Max was working on the theory that enabled Luria to do the crucial experiments on spontaneous mutations, he received a visitor in Nashville—Alfred Hershey, who had come from Washington University in St. Louis where he did research with J. Bronfenbrenner. (Bronfenbrenner was one of the first people to study phage in America.)

When Max had written up the fluctuation theory, he attached a letter to the manuscript he sent Luria in February 1943. Here he described his first impressions of Hershey: "Drinks whiskey but not tea. Simple and to the point. Likes living in a sailboat for three months, likes independence."

Alfred Hershey had been attracted by Max and Ellis's analysis of the growth of bacteriophage. In the early 1940s, he found himself in an unfortunate position. He had been more or less hired to prove what Max called Bronfenbrenner's party line: that phage were small molecules, just like normal protein molecules. Hershey's experiments were designed to show that what had been observed as large phage particles were artifacts in which the small "phage molecules" had just been absorbed by some larger entities. In Max's words, this kind of scientific work, which involved precipitating phage with antisera, made "Hershey's life unhappy for a number of years" (I).

Max contacted him in late 1942, inviting Hershey to come down to Nashville to speak about his work. As in Berlin in the early 1930s, Max had organized a very informal group of one or two dozen people from all of the science departments. They arranged talks "intended to be informative of what is going on in the various fields of science," as Max wrote to Hershey, who wanted to know how to prepare for his talk. Max answered on January 6, 1943, with the following statement: "The speaker should assume complete ignorance and infinite intelligence on the part of the audience."

Hershey's talk to Max's science club on January 1943 was entitled "Immunological Reactions of Bacteriophage." Hershey had turned down an invitation to stay at Max's house: "I have my heart set on lying all day in a hotel bed ignoring clocks and keeping you up at night with circular discourse." But he reciprocated by inviting Max to St. Louis. Before Max was able to make it there in the spring of 1943, they exchanged quite a few letters, without mentioning the historical analysis of bacterial mutations. What they did discuss in the letters was the physical chemistry of the precipitation of phage with antibodies (serum). Max later called these efforts by Hershey "his flight from reality" (I). Hershey was then not on his own, but had to work to support Bronfenbrenner's hypothesis. Only later, when Hershey began mutation and recombination studies and started with biochemistry, did he turn out original contributions.

In April 1943, Max and Luria traveled together to St. Louis; here for the first time the trio assembled that would share the 1969 Nobel Prize for Medicine. The occasion can be called the origin of the Phage Group, and it included a strange blend of characters: "two enemy aliens and one social misfit," as Max said later (Hershey, 1981).

Hershey called Max and Luria's visit to St. Louis "the first phage meeting." Max urged them to cooperate, inviting Hershey to return to Nashville to acquaint himself with the techniques used in Max's laboratory. He stressed in a letter on July 25, 1943, that a quick decision on Hershey's part would be necessary: "I hope you come here some time in the fall, maybe we can do some experiments together. . . . My guess is that there will be a great rush on phage work after the war, so we better settle the elementary problems now, that we may speak with authority later."

Hershey differed from Max, who could ignore countless difficulties to drive his research down the envisaged path. Hershey felt he could not really start to work with phage unless he knew exactly what such a bacterial virus was like. He felt torn between thinking small (Bronfenbrenner's party line) and thinking big (Max's party line). In early October 1943, he concluded in a letter to Max that "science is very discouraging," since in this case, "experiments will always be indecisive."

Max answered immediately, extending an invitation and assuring Hershey he would be welcome at any time. Beginning in Latin, Max tried to help Hershey overcome his depression: "Sicut cervus desiderat at fontes aquarum, ita anima mea desirat at te, Alexander Dembrowkowitsch Hershey. . . . I hope you will recover from your Faustian pessimism. . . . You have lived too long in the depressing atmosphere of piddling bacteriologists." For Max, the question of phage size was indeed settled—"for many reasons." His main argument for this conclusion, aside from experimental evidence, was that "the large phage makes things simple."

Max was right here, and he succeeded in convincing Hershey. Hershey himself remained appreciative of Max's generous concern about his state of mind and his scientific progress. About forty years later, in a talk in August 1981 during the dedication ceremony for the Max Delbrück Laboratory at Cold Spring Harbor, Hershey described Max's leading role in the history of molecular biology, which in recent years had moved so fast that the two scientists "tended to feel that we have been present at the creation." Hershey pointed out that Max "kept his eye on the big questions, even before they could be put into words. That is something few scientists can do. [And] . . . he devoted great effort and intelligence to encouraging, appreciating, and steering the work of others, probably often at the expense of his own. That generosity of spirit was one of his chief characteristics" (Hershey, 1981).

THE TRANSFORMING PRINCIPLE

In the same talk at Cold Spring Harbor, Hershey also indicated why he remembered the "first phage meeting" in St. Louis so vividly: "I can fix the date (April 1943) because [Max and Luria] brought the news of Avery's experiment on transformation of pneumococcus" (Hershey, 1981).

Oswald Avery, Colin MacLeod, and Maclyn McCarthy at Rockefeller University had provided the first experimental evidence that at least *some* genetic information is stored in macromolecules that were identified as nucleic acids (DNA) (Avery, et al., 1944). Working with the bacterium *Streptococcus pneumoniae* that causes human pneumonia, F. Griffith had isolated a variant that no longer was pathogenic. Colonies of this mutant no longer looked smooth like the pathogenic (S) strain, but appeared of rather rough (R) morphology. While injecting mice with S strain would kill the animals, injecting with R strain would not. The most important observation was made when heat-killed S strain was injected along with nonpathogenic R strain. The combination would kill the mice. It was concluded that a dead bacterium could transform a living variant.

The problem was to identify the chemical nature of what Avery and his coworkers called the transforming principle. In 1944, they concluded that it is DNA, thereby making what Gunther Stent called a premature discovery (Stent, 1978). At that time, DNA was poorly characterized; it was believed to be merely a tetranucleotide and thus not versatile enough to carry hereditary information. DNA was considered, in Max's words, "a stupid molecule." This view of DNA as a monotonous structure was put to rest around 1950 by Rollin Hotchkiss and Erwin Chargaff, with the help of new quantitative techniques. Paper chromatography allowed them to separate and estimate the constituents of nucleic acids, and later made possible the study of quantitative aspects of transformation when such transformation was discovered to occur between drug-sensitive and drug-resistant pneumococci.

Following the discovery of the transforming principle, the important role of DNA began to be more fully recognized. It was only nailed down when another great milestone in the history of phage research was erected in 1952. By that time, what Schlesinger had pointed out in 1934 had been convincingly demonstrated—that phage consists of protein and nucleic acids. Hershey could then show, in collaboration with Martha Chase, that only the phage DNA enters the bacterium; the protein remains outside at the cell surface (Hershey and Chase, 1952). They showed that *all* the genetic material of a phage was composed of nucleic acids, and that the proteins did not contain hereditary information.

It has often been argued that during this time the Delbrück group—that is, the Phage Group—did not take enough notice of Avery's experiments and thus missed essential molecular details by neglecting certain chemical considerations. This criticism misses the point. Max was interested in the biological phenomenon of replication; he wanted to see if physics could lead to a new law describing replication, just as the ideal gas law describes a gas. Max didn't care if protein or nucleic acid was replicated. Aside from the fact that both substances were poorly understood at that time, there was also no procedure available that would have enabled Max to experiment with these macromolecules.

Some people even detected a depreciation of chemistry in Max's approach to biological systems. It is true that Max never mastered any of the elementary procedures used in chemistry and biochemistry, and that he was impatient with biochemistry's fascination with metabolism, in the sense of converting one small molecule into another. Max never believed that further pursuit of such biochemistry would lead to an understanding of the nature of the gene, its replication, and its effects. But it is not true that the group of three ignored Avery's discovery. They talked about it in 1943, and on February 25, 1944, Max mentioned the paper in a letter to Hershey: "I suppose you have seen . . . Avery's paper on the identification of the mystery substance which causes type conversion in pneumococci. He believes that it is a ribonucleic acid. Mirsky seems to think that the evidence is no good, but I don't know. The paper is in the last issue of the *J. Exp. Med.*" Max's attitude toward these results was very positive: "It seems to me that Genetics is definitely loosening up and maybe we will live to see the day when we know something about inheritance in bacteria, even though the poor things have no sex."

In 1972, Max described his judgment of Avery's discovery in a letter to Gunther Stent:

In the late forties we did not talk much about DNA or Avery's discovery because it was not helpful to do so, *not* because we didn't believe it. Why was it not helpful? In the *early* part of that period because we did not see how DNA could carry *specificity*. Therefore, if one accepted Avery's result one thought of it as an enzyme induction effect . . . not as a gene transfer. In the *later* part, due to Chargaff and Hotchkiss' findings, one began to think that DNA could carry specificity, but *I* at last, thought of that specificity as tertiary structure, as in proteins, though conditioned, as in proteins, by sequence. When the Hershey-Chase experiment came along, it was accepted as evidence for the *generality* of the Avery finding, applying to *many* genes, and therefore ruling out an unspecific enzyme induction effect. The Chargaff-Hotchkiss data had, of course, prepared the ground for making this conclusion acceptable. However, even after Hershey-Chase, the shoe was still on the same foot, in the sense that DNA was just another macromolecule with awfully complicated *specific* structure. Then, a few months later, in the wake of the α-helix, came the Watson-Crick denouement forcing one to accept a *simple* structure for DNA. . . .

So, I would say, that Avery's discovery was not much talked about because it was premature, to be sure; but *logically* premature, not *psychologically*. It did appear out of context and the context had to be filled in, and was filled in, *painstakingly*, by Chargaff and Hotchkiss. I don't think that greater intelligence or openmindedness would have been helpful at the time, as long as the facts were missing.

In the original paper, O. T. Avery stated very carefully that the transforming principle might be DNA. He was much more outspoken in a letter that he sent to his brother Roy, who was in the Department of Microbiology in the Medical School at Vanderbilt University. Here he stated outright that it *was* DNA. After Roy Avery received the seventeen-page handwritten letter in the spring of 1944, he ran into Max while walking at the university and showed it to him.

Max later helped to retrieve this important letter. He had been invited to chair a symposium organized by the National Academy in Washington, D.C., to highlight the history of the discovery of DNA as an information storage molecule. Max contacted Roy Avery, who found the letter after spending a week going through old boxes. The letter would help establish Avery's conviction that DNA was the hereditary substance. Unfortunately, Max left the letter at home; he had to call Manny, who read the letter over the phone to Washington, where it was transcribed. Thus Max helped clarify Avery's position by resurrecting this original source.

FORTUNATE CHOICES

In 1944, phage genetics was still a quiet playground, wide open for all who wished to join in. Nevertheless, the few who had followed Max's invitation had brought in such a great variety of phages that the nomenclature was becoming confusing. Since Max expected even more scientists to turn to phage when the war came to an end, he decided to act. Accordingly, he arranged what is sometimes called the phage treaty of 1944. In the summer of 1944, the phage workers under Max's influence made an important decision. Since every investigator had his own private collection of phages and host bacteria, it was almost futile to compare results from different laboratories or even to gather enough information about one system. Max insisted that researchers concentrate on a set of seven phages active on the same host, the now famous *E. coli* strain B and its mutants. The set was called T1, T2, . . . T7 (T being the abbreviation for type). These seven phages were chosen because they gave easily countable plaques, and because the phage-resistant strains of *E. coli* B could be freed of the phage to which they were resistant (Demerec and Fano, 1945). These seven phages were thus "well behaved." Morphologically they can be classified into four groups, the best known of which consists of T2, T4, and T6. These are sometimes referred to as T-even phages.

Once this decision was made, Max held to it firmly. He would not look at results of experiments done with strains other than those included in the treaty. Though Max never found this easy, he felt a good researcher must tame his curiosity.

All seven phages included in the treaty infect and ultimately lyse their host cell. Called virulent phages, they must be distinguished from those bacterial viruses that, after entering a bacterium, do not multiply but rather insert their genetic material into the host's chromosome. These are called temperate phages; their bacterial host is not lysed but survives as a carrier of the phage, becoming what is called a lysogenic strain. When lysogenic strains of *E. coli* were discovered in the early 1920s, they were misinterpreted as bacteria contaminated with phage.

From his earliest entry to the phage world, Max worked with virulent phages. When he collaborated with Luria in 1943, they used the virulent phage T1. This was a fortunate choice for the fluctuation paper since it is now known that lysogenic bacteria are immune to infection by bacterial viruses that are of the same type as their prophage. Had they applied a temperate phage, some bacteria would have become lysogenic and thus "resistant." In other words, Max and Luria would have had to conclude that the bacterial "variants" acquire their resistant character by contact with the phage; this incorrect conclusion would have added credence to the Lamarckian view.

WHAT IS A GENE?

About the time the role of DNA was discovered and Max was corresponding and collaborating with Hershey and Luria, he complained frequently that his teaching duties left too little time for research. He needed time to prepare a series of lectures he was to give at Vanderbilt's School of Medicine in April and May 1944 entitled "Problems of Modern Biology in Relation to Atomic Physics." These presentations were reproduced without change in February 1946 and distributed to the grantees of the American Cancer Society; they remained unpublished otherwise.

After describing his view of the limitations of applying atomic physics to biology, Max proceeded in these lectures to try to determine how far atomic physics could help in understanding living cells. He discussed metabolism; photosynthesis; genes, their actions and mutations; and bacterial viruses. After first dealing with what he called ordinary biochemistry ("what the cell does with substrates introduced from without"), Max introduced genetics by reviewing the steps that lead to the picture of the gene as a material particle. He described the importance of Mendel's experiments, which showed there are factors that segregate and that can be recovered unchanged from a variety of phenotypes. He explained that cytologists were able to localize these factors in sequence along chromosomes. Then he speculated about the outlook of bacterial genetics:

> We owe our present knowledge, as it were, to two lucky accidents, the bisexuality of the higher organisms and the occurrence of crossing over at meiosis. Since our knowledge of the gene is so intimately tied to those two features, we may wonder whether genes exist also in organisms where there is no sex and no crossing over, as in bacteria.
>
> It remains to be seen whether the lack of sex and of crossing over in bacteria is merely an handicap for the study of the gene in this group.

Max pointed out that genetics now required a biochemical approach: geneticists had produced "three magnificent problems" for biochemists:

1. What do genes consist of?
2. How do they reproduce?
3. How do they act?

Max thus publicly stated his appreciation of biochemistry about a month before he learned of Avery's transforming principle. His optimism as to biochemistry's potential to solve the mysteries of genetics was influenced by the tremendous breakthrough that had been achieved by George Beadle and Edward Tatum in 1941 while working with *Neurospora*.

Drosophila geneticists at this time were attempting to work out the chemistry of known genetic differences. When Beadle and Tatum reversed this procedure, selecting mutants in which known chemical reactions were blocked, they discovered an intimate relation between genes and proteins—a relation known as the one-gene-one-enzyme hypothesis (Beadle, 1966, in PATOOMB). They found that if a single gene was mutated, a particular reaction could not occur within *Neurospora* because that gene product, the enzyme that catalyzed the reaction, was not functioning. They reported their results at the 1940 American Association for the Advancement of Science meeting in Dallas, Texas.

Max heard about this advance in biochemical genetics through his friend Norman Horowitz, who had attended Beadle's seminar at Caltech in September 1941. Horowitz remembers the silence after Beadle's short presentation of only thirty minutes. All in the room were stunned. Finally Frits Went rose, thanking the speaker for this proof that biology was not a dead subject.

The Rise of
Molecular Genetics

PHAGE IN COLD SPRING HARBOR

We spent the war in Nashville in a fabulously peaceful way. For me as
an "enemy alien" doing research for military purpose was out of the
question. I had to teach for the army the basics of physics, as for a while
almost all ten million soldiers were supposed to learn physics. This
enthusiasm for science soon cooled off. When I arrived in this country,
I became an experimental scientist, working with cute little viruses, that
attack bacteria. Fortunately, these experiments don't require much
money and nobody figured they might be useful for the war. (For the
next war chances are less favorable in that respect.) That way I really
lived a sheltered life. This is a thing of the past now. All of a sudden the
whole world starts to take an interest in my viruses. . . . We don't feel
overpowered as yet.

Max wrote this letter from Cold Spring Harbor to Jeanne Mammen
shortly after World War II ended in 1945. In that year, Max himself
had taken the step that was to set off a rapid increase in the number
of people working on phage. He had organized the first summer
phage course at Cold Spring Harbor; such a course would become
an annual event for more than a quarter of a century.

Immediately after they had started their collaboration, Max and
Luria sought for a way to recruit enough students to maintain the
momentum created by their initial discoveries. For a few years,
they had conducted experiments together every summer in Cold
Spring Harbor, starting in 1941. Max recalled that in the fourth
summer, Luria suggested that they bring people into phage by giving the summer course. When Max assumed the responsibility for
organizing the first phage course, he took a decisive step in the
foundation of molecular genetics. The course ran for twenty-six

years; its alumni included many of the prominent figures in the new biology.

The phage courses at Cold Spring Harbor mark the onset of biology as an exact science. The science of life became connected to the well-defined branches of physics and chemistry. From this point on, Max would distinguish Cold Spring Harbor "graduates" from those who had "never taken the phage course." In a more contemptuous way, he would also describe the old method of doing biology as "stamp-collecting."

Why did Cold Spring Harbor, "a small station of meager resources on the shores of Long Island Sound" (PATOOMB, 1966), become a breeding ground for molecular geneticists? One reason exists in the person of Milislav Demerec, director of Cold Spring Harbor Laboratory from 1941 until his retirement in 1960. Demerec had changed his research interest in the early forties from *Drosophila* to bacteria and their viruses when he realized they were the better choice for analyzing mutations. He also wanted others to take this step. Thus when the idea of Max and Luria's summer course was suggested to him, he helped make it work.

Max favored Cold Spring Harbor for several reasons. First, it is a small place, readily allowing the integration of personal and scientific life (Fig. 21). It is only a short walk between the motel on the grounds and the laboratory rooms. On the way, one crosses a small

21. *View of Cold Spring Harbor Laboratory buildings. Reprinted by permission of Cold Springs Harbor Laboratory.*

road leading to the beach and the tennis court. This was a perfect setting for Max. He loved playing tennis; the advantage at Cold Spring Harbor was that a tennis game also provided a marvelous opportunity for scientific discussion. While strolling to the court and back, Max would learn about the research activities of others; he would also offer suggestions as to who his partners might contact to advance their efforts. If there was no science to discuss further, Max would sometimes pose a statistical problem. For instance, he asked how one could change the way points are counted in tennis in order reliably to know, in the shortest possible time, who is the better player. Max had always been frustrated by the system of scoring in tennis, but his question led to intricate formulas. No one came up with an easy way to change the method of scoring.

During mealtimes, Max would talk to those who didn't play tennis. At Cold Spring Harbor, everyone takes his meals in a common cafeteria. Max would never sit at an empty table; he always looked for people to talk to.

By American standards, research and teaching at Cold Spring Harbor has a long history. The place was founded last century, and greatly extended in 1921 when the Carnegie Institution installed a Genetics Department there. In 1928, the Cold Spring Harbor Laboratory for Quantitative Biology was established by the physicist Hugo Frick. It epitomized a spirit of interdisciplinary cooperation. Starting in 1933, an annual spring symposium was organized with financial help from the Rockefeller Foundation. Each symposium dealt with an aspect of quantitative biology, and the proceedings were later published as monographs.

Max participated in the 1941 symposium, entitled "Genes and Chromosomes: Structure and Organization." Here he communicated his theory of autocatalysis (D 19). The 1946 symposium on "Heredity and Variation in Microorganisms," in which Max played an instrumental part, became the most memorable meeting in the history of molecular genetics. Here the one-gene-one-enzyme hypothesis triumphed and the discovery of genetic recombination in bacteria was announced.

THE PHAGE COURSE

Cold Spring Harbor has been known for its excellent courses since 1927, when "General Physiology" was first offered at the laboratory. The 1929 Annual Report's description of the summer activities could apply just as well to one of the postwar phage courses: "The Laboratory is assured of a serious group of students, large enough to form a very appreciable nucleus for the future body of biologists in this country and small enough to be easily assimilated into the life of a research institution." Most of the early courses at Cold Spring Harbor were maintained into the 1940s. The Depression did not halt the summer teaching, mainly due to the faith and support of the Long Island Biological Association.

The first course on bacteriophage, in 1945, was given in the same laboratory building that had housed Max and Luria's summer research during the early 1940s. Originally built in 1925, it underwent a major renovation fifty years later; in 1980, a new annex was constructed. The laboratory has now been renamed in honor of Max, to recognize his forty-year association with Cold Spring Harbor.

The goal of the courses on bacteriophage was to make phage an important line of research and to develop its potentialities. Though only a few who took the course actually became phage workers, the summer courses did help toward the goal, since "we recruited a number of people who could read the phage literature with understanding," as Max put it (I). The advertisement for the course read as follows:

> Bacteriophages exhibit many of the properties of viruses and can serve as models for the study of fundamental problems in virus research. The bacteriophages are also a useful tool for the study of mutations in bacteria.
>
> Since 1941, research on bacteriophages has been carried on during the summer months by several investigators (Cordts, Delbrück, Gest, Luria, Spizizen) at the Biological Laboratory. Moreover, since 1943, work on bacteriophages has also been carried on at the Department of Genetics of the Carnegie Institution (Demerec and Fano).
>
> While this work has aroused some interest among biologists, the number of investigators actually involved and intimately acquainted with the techniques of such work is exceedingly limited. It seems worthwhile and appropriate, therefore, to offer at the Laboratory a short but intensive laboratory course in the techniques and problems of this field.
>
> The purpose of the course is to acquaint the student with some of the techniques used in bacteriophage research, and with recent results of such work.
>
> The course will consist of about nine half-day laboratory periods and nine half-day study periods for the evaluation and discussion of the

experiments performed by the students. No outside work should be planned by the student, since it may be expected that his full time will be occupied by this course.

The first course was unusual in the sense that Max set prerequisites: "Facility in the processes of multiplication and division of large numbers." An admissions test was given. Max wanted to be sure that the participants in the course could handle the appropriate dilutions properly.

Max planned the course so that the students would learn in nine days what he had produced in the eight years between 1937 to 1945. The famous one-step-growth experiment was scheduled for the sixth day; it was introduced by the following description:

ONE STEP GROWTH EXPERIMENT

Virus growth occurs in this manner: a virus particle gets attached to a bacterium and grows within the bacterium for a certain time. The bacterium then bursts and a large number of virus particles are liberated.

Problem

(a) Determine the *latent period* of virus growth, i.e., the period from adsorption to burst.

(b) Determine the *average burst size*, i.e., the average number of virus particles liberated from one bacterium when it bursts.

Procedure

Mix about 108 bacteria per cc with about 107 virus particles per cc. This will be the *adsorption mixture*. Allow four minutes for adsorption. Then dilute 1:1000. This will be the *first growth tube*. Dilute from this further 1:100. This will be the *second growth tube*. Plate samples from first and second growth tubes, diluting them 1:10 with plating bacteria.

When Max first saw phage plaques in the basement at Caltech, he had never touched a pipette before. Like the phage students Max envisaged for the first course, he had never done a dilution series or prepared a solution. His advantage had been that he was mathematically prepared, ready to do quantitative experiments. Max planned his courses for students with similar backgrounds. He didn't want to have to teach them calculus; they would have to come prepared.

In Nashville in early 1945, Max practiced the course with two Vanderbilt students. These two, Milton Bush and Norm Underwood, were the first to retrace Max's path into the phage world.

The first summer course on bacteriophage at Cold Spring Harbor ran from July 23 to August 11, 1945. Max was the instructor; his assistants were A. H. Doermann and J. Reynolds, and the class consisted of six students: M. Baylor from the University of Illinois, R. D. Hotchkiss from Rockefeller University, H. M. Kalckar from the New York City Public Health Research Institute, P. Margaretten of the University of Illinois, S. Mudd of the University of Pennsylvania Medical School, and T. Sigurgeirsson of the University of Reykjavik, Iceland, who was sent by Niels Bohr. Max taught the course for the first three summers. Mark Adams, from New York University's College of Medicine, helped in 1947 and took over in 1948. By that time, the course lasted three weeks and was accompanied by official seminars, the first of which was given by Luria.

Aaron Novick, who took the phage course in 1947, has described Max's method of bringing physicists into biology. The course introduced the participants "to a biology that had been made comfortable with backgrounds in the physical sciences. In that course we were given a set of clear definitions, a set of experimental techniques and the spirit of trying to clarify and understand. It seemed to us that Delbrück had created, almost singlehandedly, an area in which we could work, and after the . . . course we felt ready to embark on our own without further preparation" (Novick, 1966, in PATOOMB).

WHAT IS LIFE?

Max's course received some unexpected publicity through the attention given to Max in Schrödinger's *What Is Life?* The book brought Max renown, giving a boost to the recruitment of phage workers. Schrödinger, a founding father of quantum theory, popularized Max's quantum mechanical analysis of the gene. Terming Max's 1935 theory of the stability of the gene the "Delbrück model," Schrödinger pointed out that "there is no alternative to the molecular explanation of the hereditary substance. . . . If the Delbrück picture should fail, we would have to give up further attempts."

This attention was unexpected for Max. In a letter to the science historian R.C. Olby, he mentioned that he had known Schrödinger personally in Berlin:

> He was still very much there when I came to work with Lise Meitner in the fall of 1932 and I saw him a number of times and was quite friendly with him. However this was before my collaboration with Timoféeff. I do not think that Schrödinger met either Timoféeff or Zimmer. I am not certain how he came to take an interest in our paper. I presume that I sent him a reprint when the paper appeared in 1935, I believe, to wherever he was, either at Oxford or at Dublin. But of course his book was ten years later and I was not in contact with him during these subsequent years, so I do not know at what time he chanced to pick it up and read it.
>
> The Schrödingers must still have been in Berlin *at least* till February, 1933, because I remember a costume party in their apartment about that time, for which I borrowed the uniform of the porter at the Harnack-haus and acted the part of their butler at their party. When the news of Schrödinger's Nobel Prize came in October, 1933 I wrote a letter of congratulations to Oxford, pretending to be his old butler Schulze, and received a reply to "My dear Schulze . . . remembering your many years of faithful service I am giving you an annuity of one hundred pounds (of potatoes)." (Olby, 1974)

What Is Life? is based on lectures on the physical aspect of the living cell that Schrödinger delivered at Trinity College in Dublin in 1943. In these talks, he raised the following question:

> How can the events in space and time which take place within the spatial boundary of a living organism be accounted for by physics and chemistry?

He gave this summary answer:

> The obvious inability of present-day physics and chemistry to account for such events is no reason at all for doubting that they can be accounted for by those sciences.

Schrödinger, like Bohr more than a decade earlier, discussed only questions of great generality. One such question was how organisms can overcome the Second Law of Thermodynamics which claims an increase of disorder. These all too quickly become philosophical problems. The book was published at a time when biology was being removed from the hands of traditional biologists. Physicists moved in, changing the questions being asked about biological problems and the nature of acceptable answers (Yoxen, 1979).

Unlike others, Max went beyond these general considerations; he turned to quantitative experimentation and tried to answer the questions that were being formulated by others. For that, he gained admiration that enabled him to play an influential role. The influence of Schrödinger's book is discussed in detail in the literature (Olby, 1974; Yoxen, 1979). It is known that four outstanding biologists who read it were attracted to the mysteries of biology as a result: Seymour Benzer, Francis Crick, Gunther Stent, and James Watson. The reasons they offered for the attraction vary from personal to philosophical ones, as they are described in the 1966 Delbrück Festschrift. But all agree that *What Is Life?* focused their attention on Max, making him the natural father figure of the emergent Phage Group.

THE PHAGE GROUP

The Phage Group consisted of scientists who addressed themselves to the kind of genetics problems Max and Luria were interested in. The group came into existence during the informal meeting of Max, Luria, and Hershey in St. Louis in the spring of 1943. The first official meeting was organized in March 1947 in Nashville, Tennessee. It attracted eight people: M. H. Adams, T. F. Anderson, S. S. Cohen, M. Delbrück, A. H. Doermann, A. D. Hershey, S. E. Luria, and M. Zelle. Starting in 1949, Leo Szilard conducted a monthly Midwest phage meeting, with James Watson and Luria among the regular participants. From 1950 on, the annual phage meetings at Cold Spring Harbor attracted so many people that, in one participant's words, "the growth rate of molecular biologists . . . was much higher than that of humanity" (Luria, 1966, in PATOOMB).

What brought Max renown was his leadership of the Phage Group. The two locations at which he based his activities—Cold Spring Harbor and Caltech—became the Mecca and Medina of the phage world. In creating this group, Max used as a model the group around Niels Bohr that created quantum physics. The essence of the Phage Group was that "it was open and cooperative in strict imitation of the Copenhagen spirit in physics," as Max wrote Robert Olby in 1972 (Olby, 1974). It was Max's great achievement "to keep the Phage Group pure" (A. Hershey) and to induce its members to treat each other with "joyful irreverence" *(fröhlicher Respektlosigkeit),* as Max described the attitude he had admired in Copenhagen. Max had learned in Copenhagen, from the Russians and particularly from George Gamow, that teamwork can exist only if accompanied by frank human relations (Fig. 22).

A group formed around a common scientific interest still needs a leader who can act as a court of appeal. Max took it as his mission to sort out any bad science; he created an atmosphere in which criticism was meted out ruthlessly, no holds barred (Olby, 1974).

22. *The Phage Group at Caltech (1949). Left to right: Jean Weigle, Ole Maaloe, Elie Wollman, Gunther Stent, Max, and G. Solti. Reprinted by permission of Cold Spring Harbor Laboratory.*

But he never ignored fun. Social activities and close personal relationships were encouraged—such a group could work well only if kept to a small size.

One prominent quality of Max's special style of leadership was his devastating criticism. He would interrupt seminars, calling out: "I don't understand a word you are saying. Start all over again." Max would challenge those who felt they had presented conclusive evidence by announcing he didn't believe a word. This brusque manner has been commented on by many. According to Ray Owen, Max's colleague at Caltech, "sometimes his behavior seemed inhumane, because he valued an impersonal search for truth, and held a standard that permitted no sham or sloppiness to go unmasked. But he had an extraordinarily warm and humane and perceptive heart. A sense of humor pervaded all of his relationships" (*New Scientist,* 119, [1981], p. 169).

Certainly Max's presence in seminars, though eliciting mixed feelings in the speaker, encouraged a meaningful discussion. His rough handling of speakers continued the tradition established by Bohr, who used to preface his interruptions with "Only in order to learn." Although Max interfered in seminar presentations as frequently as Bohr did, he was considered the more polite of the two. Yet he felt, as Bohr did, that one of the most important activities of science was to ask questions. His most influential questioning came in Cold Spring Harbor in 1946, at the first symposium held after World War II.

THE EXPLOSION OF GENETICS

The 1946 Symposium on Quantitative Biology was dedicated to "Heredity and Variation in Microorganisms." It became a memorable event in the history of molecular genetics, for here biologists first learned of the discovery of genetic recombination in *E. coli* and in phage. Thus, it became clear that even lowly bacteria and their viruses possessed genes, and that they were capable of genomic reshuffling in natural populations. This had been considered an exclusive property of those organisms that had progressed from the vegetative to the sexual mode of reproduction.

Genetic recombination in bacteria was announced by Joshua Lederberg and Edward Tatum. They had discovered this phenomenon when they found bacteria which combined certain traits of two parental strains that had been mixed (Lederberg and Tatum, 1946 a, b). The fact that phages also could exchange material had first been observed by Hershey, and independently, in Max's laboratory in Nashville (Hershey, 1946, and D 35). The road to this discovery was paved by Hershey's isolation of a phage variant that appeared spontaneously on his plates. He had noticed, among the normal plaques which were small and surrounded by a turbid halo, different ones that were larger, with a sharp edge. Hershey isolated this variant particle, finding that its progeny produced plaques identical to each other in morphology. This allowed him to call this strain a phage mutant. It was named T2r, since it was a mutant of T2 phage that showed rapid lysis (disintegration).

When Hershey infected *E. coli* with T2 and T2r, both phages could produce progeny. Max considered this only as "an exceptional case in which the mutual exclusion mechanism appears to function only partially," as he wrote the Rockefeller Foundation in June 1945. But Hershey did not accept this interpretation. His pursuit of the problem revealed an effect complementary to mutual exclusion, namely recombination. Hershey arrived at this discovery after he isolated a mutant of phage T4 that also exhibited r-type morphology. When he then infected bacterial cells with T2 and T4r, he found among the progeny not only T2 and T4r, but also new recombinant phages: T2r and T4.

The fact that bacterial viruses have a sexual form of reproduction and thereby violate the principle of mutual exclusion was also noticed at Vanderbilt. It came about in the following way: W. T. Bailey, Jr., and Max were working with two phages that had two distinguishing characteristics but could reproduce in the same bacterium, *E. coli* strain B. Phage T2 produced small plaques and could also prey on a variant (a mutant) of *E. coli* B, called A. The other phage, T4r, produced large colonies and could destroy still another mutant of *E. coli* B, called C. When T2 and T4 were added to a bacterium, viruses of both these parent types were released upon burst, as expected. But in addition, two new types of viruses were released with their characteristics switched. One of the new types produced large colonies and could destroy A; it was thus called T2r. The other produced a small plaque and could destroy C; it was thus called T4. In other words, somewhere inside the bacteria during reproduction the parent viruses had exchanged genetic material. Of course, the process was not called exchange of genes then, for no one knew for certain if genes and genetic material were present in phages. Max's description put it this way: "the parents had got together and exchanged something" (D 37). He did not follow up on the finding.

In a sense, Max was disappointed in the recombination finding, for it complicated the simple situation he wanted to analyze. Instead of being an elementary and basic fact of life, reproduction turned out to involve complicated exchange mechanisms. Ever since Max had begun to analyze phage multiplication, the system had exhibited new complications at every turn. The first blow was dealt by the discovery of interference (D 20, D 21, D 31). Instead of getting a direct look at the multiplication process as they had hoped, Max and Luria found strange exclusion effects. Max attempted to save simplicity by a "key enzyme" hypothesis or "penetration barrier" model, but neither approach worked.

The terms that excited Max the most then—"interference" and "mutual exclusion"—are hard to find in modern genetics textbooks. (Interference is now attributable to a phage-induced change in the bacterial envelope.) But despite Max's distrust, recombination turned out to be the direct path to phage success. Soon after the 1946 Cold Spring Harbor Symposium, Hershey and Raquel Rotman began to construct a genetic map of phage T2, using the reasoning first applied by Morgan and Sturtevant twenty-five years earlier in their work on *Drosophila* genetics.

MUTATIONS IN PHAGE

The phage mutants that allowed the discovery of recombination were of two types. Hershey had discovered the plaque-type mutants which produce plaques of different morphology. The other set, discovered by Luria, was called "host-range" mutations (Luria, 1945).

Luria had begun to look for phage mutants immediately after he and Max had settled the question of the spontaneous nature of the appearance of resistant bacterial variants. Their joint experiments led to an obvious conclusion: If bacteria can mutate to become phage-resistant, then phage undergo a mutation, becoming virulent again. Otherwise, all sensitive bacterial cells would have been destroyed long ago; the viruses would have been deprived of suitable hosts, to vanish altogether.

As predicted, Luria discovered "host-range mutants"—phage that are able to get around the bacterial resistance and infect the cells yet again. If *E. coli* now would respond with a "super-resistant" mutation, phage would answer with another variant that would circumvent the newly laid barriers. Thus, a delicate mutational equilibrium allows the coexistence in nature of bacteria and their viruses. Such an equilibrium has to involve a balance of the genetic material, but the term "equilibrium" only makes sense if a system is in a dynamic, as opposed to a static, state. So the genetic material can be seen from this result to be in dynamic equilibrium. In other words, just by looking at the coexistence of *E. coli* and phage, one could realize that the description of genes as a solid—Schrödinger introduced the notion of an "aperiodic crystal" (1945)—is not sufficient. Genes show some characteristics of fluids, for they are a moving and movable system.

CONDITIONAL LETHALS

The highlight of the 1946 symposium was the triumph of George Beadle and Edward Tatum's one-gene-one-enzyme hypothesis. Several presentations showed that one could restore growth of auxotrophic mutants in fungi by adding a single metabolite to the medium. This was interpreted—in strong support of the theory—as indicating that the mutant lacked the enzyme to catalyze the synthesis of the metabolite, since the gene responsible for it had been mutated.

The one-gene-one-enzyme hypothesis, constituting a biochemical definition of the gene, had first been presented five years earlier. Now in 1946, it was celebrated. Only Max remained somewhat reserved. After a presentation by Beadle's collaborator, David Bonner, on mutant strains of *Neurospora,* Max rose to point out that these mutants fail to prove the theory, since mutations in genes controlling many enzymes would not have been found by the methods that were employed to isolate the mutant strains. Max challenged the proponents of the hypothesis to devise experiments by means of which the theory could be disproved. The following transcription illustrates his inquiry about the conclusiveness of the evidence:

Delbrück: It has been pointed out by the speaker that the evidence accumulated in the Neurospora work is compatible with the assumption that there exists a one-to-one correlation between genes and the various species of enzymes found in the cell; and the hypothesis is advanced that quite generally the function of genes consists in imparting the final specificity to enzymes. The question arises whether the evidence obtained actually supports the thesis, beyond the mere fact that it is compatible with it. The possibility exists, and should be discussed, that the experimental approach applied in the Neurospora work is such that incompatibilities would be unlikely to arise even if the thesis were not true at all.

Max went on to show how "the thesis might be untrue." He constructed a case that could contradict the hypothesis. Then he summed up his argument:

In order to make a fair appraisal of the present status of the thesis of a one-to-one correlation between genes and species of enzymes, it is necessary to begin with a discussion of methods by which the thesis could be *dis*proved. If such methods are not readily available, then the mass of "compatible" evidence carries no weight whatsoever in supporting the thesis.

In his answer, David Bonner pointed out that the "one-one concept" had evolved from observing that the loss of the ability to carry specific reactions was associated with the loss or alteration of a single gene. (The "gene" in this sentence is a unit defined by genetical experiments, that is, by studying the inheritance of mutations and their complementation.) The concept thus does not imply that a single gene controls the entire production of a single enzyme; Bonner assumed that "probably many genes are necessary in the production of the required polypeptide structure." But he disagreed with Max's summary, since the available data could all be accounted for on a one-one basis. He felt "it would be of little value to devise a more complex thesis of the relation of genes to enzymes."

Max, who indeed favored ideas that revealed simplicity, objected because there was no way to falsify the hypothesis. Here Max presented the same argument that Karl Popper had advanced in his *Logic of Scientific Discovery,* originally published in German in 1934. Popper had pointed out that "theories are . . . *never* empirically verifiable"; the logical focus of science should be such that *"it must be possible for an empirical scientific system to be refuted by experience."* If a hypothesis is verified by an experiment, then nothing is gained (and nothing lost either). Only falsification of a theory allows science to advance to a better theory.

Challenged by Max's demand for falsification, Norman Horowitz went on to exploit the enormously powerful method of isolating temperature-sensitive mutants to try to disprove the one-gene-one-enzyme hypothesis. Horowitz's response to Max eventually revolutionized the study of physiological genetics.

In his paper presented at the 1951 Cold Spring Harbor Symposium on "Genes and Mutations," Horowitz introduced the term "indispensable function." A mutant has lost such a function if it is incapable of growing on a complete medium. He then restated Max's question as to whether incompatibilities with the one-gene-one-enzyme hypothesis could be detected even if they occurred: "The point of Delbrück's argument was that if any gene has more than one primary function, it is likely that at least one of these is an indispensable function; in which case mutation of the gene would not be detected" (Horowitz and Leupold, 1951).

The way to clarify this dilemma is to find out what percent of gene functions are indispensable, and thus what the chances are for recovering a mutation of a gene with several primary functions. But the question of how to determine the proportion of indispensable functions "would seem almost by definition to be unknowable, in which case the one-gene-one-enzyme idea must be banished to the purgatory of untestable hypotheses, along with the proposition that a blue unicorn lives on the other side of the moon" (Horowitz and Leupold, 1951).

There was, however, a basis for comparing the frequency of mutations that result in a loss of an indispensable function to that of mutations which cause loss of a dispensable function. This method uses so-called temperature-sensitive mutants that show their genetic defects only at particular temperatures. For example, some mutants require a growth factor at 35 degrees Centigrade, but grow in its absence at 25 degrees Centigrade. Since the mutant strain would die without a growth factor at the high temperature, it is called a conditional lethal. Its mutation results in the loss of vital function under conditions controlled by the experimenter.

Such temperature-conditional variants of *Neurospora* were first obtained in Beadle's laboratory at Stanford in the 1940s (Mitchell and Houlahan, 1946).

After Max's objection, it occurred to Horowitz that by means of such mutations one could, in principle, recover mutants that would be lethal, and therefore undetectable, in their fully active form. If these mutants, which had simply been selected for their inability to grow at the restrictive temperature, also showed only a single functional defect, then Max's criticism was not valid.

Horowitz described the result of his analysis at a meeting on "The Origins of Biochemistry" organized by the New York Academy of Science:

> . . . by placing the mutants at the temperature at which their mutant character is manifested, it could be determined what fraction of them failed to grow when supplied with the standard complete medium used in the standard selection procedure. When I applied this test to the known temperature-conditional mutants of *Neurospora* in 1950, I could find little evidence for selection of the kind Delbrück had postulated. Leupold and I then examined a much larger number of temperature-conditional mutants in *E. coli* and found even less evidence for selection against multifunctional losses than in *Neurospora*. This result was very reassuring for the one gene-one enzyme theory. (Horowitz, 1979)

Horowitz and Leupold even used Poisson distribution to conclude that at least the majority of genes controlling biosynthetic reactions in *Neurospora* are unifunctional.

Horowitz's proof shows how complicated it can be to show that living matter operates on simple principles. The one-gene-one-enzyme hypothesis is simple and elegant (and true); the experiments that supply the evidence are ugly and complicated. No wonder hardly anyone paid attention to Horowitz's work; not even Max took much note. Max had raised a logically valid objection in 1946, then lost interest in this matter once he began concentrating on reproduction again. Max was unwilling to think about a problem on a part-time basis—he would do things well or not at all. Thus he paid little heed to conditional lethals when Horowitz and Leupold exploited this principle to meet Max's objection.

As a consequence, no member of the Phage Group thought of extending the notion of temperature-sensitive phenotypes to their

T phages. Only in 1964 did R. S. Edgar succeed in isolating and characterizing such mutants of the T4 phage (Edgar and Lielausis, 1964). Edgar writes, "From an historical point of view . . . [this] is an embarrassing parallel (in miniature) to the rediscovery of the Mendelian laws. Our notions concerning the nature of conditional lethals and their application arose in an historical vacuum. . . ." (Edgar, 1966, in PATOOMB).

One should not forget, however, that around 1950 the molecular basis of mutations was quite unclear, and there was no way of telling if it was the gene or the enzyme that was temperature-sensitive. It was unfortunate, though, that the Phage Group looked the other way when Horowitz presented his results, since geneticists "were desperately in need of good markers for formal genetics analysis" (Edgar, 1966, in PATOOMB), something the conditional lethals provide.

REVIVAL OF THE DEAD

With the discovery of recombination in bacteria and in phage, it was safe to assume that these biological entities also contained genes. And when it was discovered in 1947 that light could influence these genes, Max was overwhelmed again, hoping once more that his "Light and Life" fantasy might come true. Max was confident that Luria's discovery of what he called "multiplicity reactivation" (Luria, 1947) was "the key break . . . which soon should tell us what was what" (Watson, 1966, in PATOOMB).

This discovery of multiplicity reactivation came as a sequel to an accidental observation made in Max's laboratory at Vanderbilt University, where a strange "revival of the dead" was observed. It came about in the following way: One can "kill" a bacterial virus by exposing it to ultraviolet light; that is, when the irradiated phages are added to a bacterium, the bacterium is destroyed but no new viruses issue from it. One day, W. T. Bailey, Jr., irradiated some viruses in Max's laboratory long enough so that most, but not quite all, were killed. He wanted to determine the number of survivors, expecting to find less than one hundred virus colonies in the bacterial lawn on the plate. To the contrary, he found ten times as many. When he repeated the experiment, he got the same result. The supposedly dead viruses had, in some way, come to life again.

One has to be careful about using the terms "dead" or "living" for viruses. As Max put it, "viruses seem to lie on that uncertain and perhaps unreal borderline between life and non-life" (D 37). The trick performed by the virus is that although it seems dead outside a bacterium, it does not die; it just lies quiescent and functionless until a bacterium shows up.

Max did not follow up on Bailey's observation, but Luria did. He discovered a curious fact. A bacterium infected with only one "killed" virus dies and yields nothing; a bacterium infected with two or more "killed" viruses—this is called multiplicity—bursts and yields several hundred new viruses. Luria explained this multiplicity reactivation by supposing that each virus particle is built of a number of different subunits of equal sensitivity to radiation; the integrity of each subunit is needed for reproduction (Luria, 1947). The inactivation of only one subunit of a virus would thus prevent viral reproduction within a bacterium infected with that single virus. In the case of multiple infection, however, it is likely that a different subunit is inactivated in each infecting phage particle, so that the great majority of bacteria receive at least one complete set of intact subunits which could reassemble to yield active phage. Luria supposed that the postulated subunits were independently reproduced. Now, of course, they could be interpreted as genes; recombination could be considered the mechanism whereby active vegetative phage is regenerated.

It was immediately obvious that the study of these two phenomena of genetic exchange would likely reveal how one virus produced another, and even "something of the simple facts of life" (D 38).

MAX AT FORTY

These very exciting developments in phage and bacterial genetics took place at a restless period in Max's life. Between 1945 and 1947, he hardly had a chance to sit and do what he liked best, think about science.

In 1945, when the summer course on bacteriophage began, Max could no longer retreat from the world. His family in Europe had suffered terribly during the war. Seventeen nieces and nephews had lost their fathers. Max's brother Justus had died in a Russian camp; his brother-in-law, Klaus Bonhoeffer, had been killed by the Nazis before Germany was taken over by the Allies.

Some expected Max would move to Germany in order to help out directly there. But he felt he could support his friends and relatives much more effectively by staying in the United States and sending CARE packages. He did plan to return to Germany for a visit, however, and reestablished his European contacts by mail. Max was very pleased to receive the first letters from Jeanne Mammen after a break of several years.

To further his distractions, he was running out of money for research; he had to decide whether to stay in Nashville. Max decided to play it big. Vanderbilt's graduate dean at this time, Phillip Davidson, remembers that Max suggested setting up a molecular biology department in the South, one big enough to attract somebody like Luria. Max felt that the space he occupied in Buttrick Hall was too small. To the Rockefeller Foundation, he presented another justification for his Nashville proposal. The director of the Foundation's Natural Sciences section, F. B. Hanson, noted in his diary on March 23, 1945:

> [Max] remarks that, curious as this may seem to someone else, he has fallen in love with Vanderbilt and life in the South and has no desire whatever to leave Vanderbilt so long as he can continue his researches there. He is opposed in principle to concentration of so many of our best research men on the Atlantic Coast; [he] believes it a healthy thing if institutions like Vanderbilt were enabled to retain their best men.

Max had, however, requested an amount greater than the total assets of the Natural Science Division at Vanderbilt. The request could not be funded. His love of Vanderbilt also cooled off considerably when the university became flooded with veterans who were encouraged by the government to return to school.

In late 1946, Max was looking for a new job. In a letter to Jeanne Mammen he explained why he wanted to get away from "rather boring" Nashville:

> I am still at the crossroads. Science is now very much funded by the government. That way, it develops into an industry organized by big bosses who themselves are no longer scientists or have never been. This all smacks abominably like bureaucratic patriotism. The scientist

changes with it. He no longer is the "cross of a mimosa and a porcupine" as he was in the good old times; he becomes more a cleanshaven and perfectly organized engineer. . . . At the moment I still try to save my ivory tower; maybe Manchester is the ideal location for erecting such a tower.

Earlier that year, Max had received offers from the Carnegie Institution of Washington and from the University of Manchester in England. P. M. S. Blackett from Manchester offered him the opportunity to run a biophysics department there. For Max, Manchester was indelibly connected with Niels Bohr since Bohr had discovered the quantum theory of the hydrogen atom while working at Manchester with Ernest Rutherford. Apart from this historical and sentimental connection, Max welcomed a return to Europe. He would be closer to his relatives, who needed his help. Manny Delbrück was also ready to live in Europe.

Max considered Blackett's offer seriously, giving four lectures at Manchester in June 1946. Surprisingly, Max also requested that the university provide him with a biochemical assistant: "Without a really competent biochemist on hand, not only the phage work, but the entire operations of a biophysics department would be lame." Max was planning to start in Manchester in 1947. But on December 11, 1946, he "unexpectedly" received an offer of a professorship in biology at Caltech. Immediately, he wrote to Manchester that he "could not hope for a better setting for [his] work" than Caltech.

The offer had come from George Beadle, who had been appointed chairman of Caltech's Biology Division in 1946. He wanted the institute to shift its emphasis from cytogenetics to chemical biology or molecular genetics. Max was recommended to him by Norman Horowitz, who had joined the Caltech faculty after spending a postdoctoral period with Beadle in Stanford. After conferring with Pauling, Beadle sent Max a telegram. Max accepted the Caltech offer on December 27, 1946, informing the Rockefeller Foundation on his usual postcard. He also wrote a letter to Niels Bohr that day: "I accepted a professorship at Caltech. . . . I am very happy about this, because it signals the completion of my metamorphosis into a biologist, and because I believe that Caltech in the coming years will be to biology what Manchester was to physics in the 1910's."

The letter explained what Max needed a biochemist for: he felt biochemistry might be one of the techniques that would help reveal mutually exclusive phenomena. Max knew how powerful biochemistry was, feeling that while it could not solve the riddle of life, it might well lead to it.

The best biochemist to have taken the phage course at Cold Spring Harbor was Mark Adams. In fact, he was about to be named instructor of the course. Max wanted Adams to join him at Caltech. Originally, he intended to make it a condition for his acceptance that Caltech offer Adams a faculty position. But both Pauling and Beadle astutely suggested this might not accomplish the purpose Max had in mind, namely, to strengthen the phage work, since Caltech allows every faculty member complete freedom of choice in his field of work. They recommended offering Adams a staff position.

When Adams turned down this first offer by Beadle, Max tried to convince him that Caltech was a special place to work: "The aim at Caltech is fundamental biology, quite uncontaminated by the MD [medical doctor] spirit. There is nobody at Caltech who kills 100,000 mice per annum for futile nonsense. Cooperation is real, and so is the community of interest."

In the same letter, dated January 7, 1947, Max explained why he needed someone like Adams:

I want you . . . to help in the organization of the lab. Your experience with bacteriological and chemical equipment would be very valuable. You know that I am an amateur in both these fields, and always afraid to make a fool of myself. . . . I expect that I am going to lead a double life at Caltech, that of general think-man (because I know physics and because I put my foot into various theoretical problems) and as phagologist. I do not know exactly why they hired me, I suspect for both purposes, or on the theory of double insurance, that is, if I should turn out to be a dud in one respect I might still turn out to be alright in the other. However, leading a double life means splitting your time, and I would like to be in a position to be able for weeks on end to pay no attention to the phage end. In other words, I would like you to be the director of the phage lab, with the hope that you will permit me to work in your lab when I feel like it.

Mark Adams still did not decide in favor of Caltech. He preferred to remain on the East Coast, where he was comfortable at Rockefeller University. He died, a young man, in 1949, after beginning to write a first textbook on bacteriophages.

THE CALTECH PROFESSOR

After Max had met phage for the first time in 1937, he predicted to T. H. Morgan that this creature would keep him busy for the next ten years. When Beadle, Morgan's successor at Caltech, brought Max back to Pasadena, the ten years had already passed. Indeed, Max did begin to concentrate his efforts differently. Instead of conducting phage experiments, he began to run the Phage Group. Caltech became the place "where many of the first generation of molecular geneticists took their orders," as one of them once put it (Stent, 1981).

Max had made his own contributions to the foundations of molecular biology in the early forties; he had begun to spread the phage gospel in Cold Spring Harbor back in 1945. But he made his strongest impact during the first years he spent as Professor of Biology in Pasadena. In the collection of 1966 essays honoring Max entitled *Phage and the Origins of Molecular Biology* (which Max called PA-TOOMB), many agree that it was Max's manner as leader of the Phage Group that enforced quality. They all had to work hard to convince Max, who never tired of challenging colleagues. The numerous scientists represented make it clear that it was not easy to be around Max; but they also make it clear that Max made it easier for a scientist to be a good scientist.

Frequently, Max refused to look at experimental results if there was no theory to embrace the available material. "Enough data," he would say. Those who wished to undertake experiments just because they would work met with his scorn. He forced his coworkers to consider whether an experiment would advance understanding of a significant problem. Max was never impressed simply by data, but by the theoretical framework they supported. He challenged everyone in the Phage Group to present and scrutinize the theoretical grounding.

Max was at his best when he could debate one on one. Niels Jerne related a typical experience, involving a discussion with Max about using antibodies in phage research. Jerne objected to certain hypotheses, but Max "had an argument against this, which struck me by its brilliance but which has since escaped me." This happened frequently to others. Whenever Max presented an argument, he was quite convincing, but often his colleagues were unable to remember the details of his logic. As a consequence, they would frequently spend time trying to reconstruct Max's way of thinking, even after hearing it. It was also common for phage workers to try to anticipate his response, to take care of potential objections in advance. To this day, some of his former coworkers check themselves by questioning, "What would Max think of this?"

Though Max valued personal relationships among the phage workers, his first allegiance was to scientific truth. As a leader of phage activity, he never spared seminar speakers. If a speaker presented a theory that contained a false step, Max would invariably discover it and point it out "with great accuracy and some violence." He simply refused to waste his time; he would in public advise the speaker to think further about the subject.

Max's most famous response after a seminar was to tell the speaker, "This was the worst seminar I ever heard." For those who fainted upon hearing these words, a bottle of brandy was available in Max's desk. He didn't mind if people teased him about his rudeness. When George Feher from San Diego gave a series of seminars at Caltech, he opened the first talk by pointing out that he was especially pleased to be giving two seminars, since then at least one would not be the worst seminar Max had ever heard. Max smiled.

Often, Max would walk right out of seminars still in progress, if the speakers did not get to the point. He also could make people go through "the ordeal of a five-hour seminar," if he felt the issue at hand was sufficiently important and ready for solution. Despite the brusqueness, Max was always willing to listen; he was an active listener, following all lines of reasoning, especially those involving mathematics. He would never sit in a seminar simply waiting for the final conclusions, but would monitor the arguments closely and then draw his own "take-home lesson," as he called it.

In publications, too, Max set high standards of quality. When authors complained after he rejected a third or fourth draft that they didn't have that much time to spare on writing a single paper, Max would put on "his angelic smile" (Jean Weigle) and ask if the effort hadn't improved the presentation and at the same time clarified their thinking. Few could argue back.

For the Phage Group, Max's presence was a permanent challenge; he made things move. His most famous way of turning colleagues on was simply to state, "I don't believe a word of it," as he did when he saw plaques for the first time or when he was told the genome of phage T4 was circular. Max would not argue after saying this, but would just wait for further evidence. Even his friends felt that he often overdid this skepticism, becoming somewhat dogmatic.

Among his phage coworkers, Max enforced his own research style, which meant, apart from a distaste for chemistry, an emphasis on theory over experimental data. Max was always willing to "say a little too much in interpretation," and he wanted the members of the Phage Group to exempt themselves at least one day a week from the laboratory to have a chance to conduct intensive interpretation. If the urge to perform experiments proved so strong that others did not take time to think, Max was known to lock them up to write up their interpretations of results.

Quite a few scientists suffered from Max's demanding and challenging manner; in spite of this, he managed to keep the group's spirits up. Most of those who were connected with Max remember their time in Pasadena as the best time of their lives. This was partly due to the fact that Max not only took an interest in the work of his colleagues, postdocs, or students, but he also warmly involved their families. He often included the Phage Group in adventures that made life with his family so different from the routine lives of other scientists. Seymour Benzer has described the Delbrücks' camping mania:

Sometimes Delbrück would proclaim Wednesday and Thursday as a weekend, to avoid crowds and highway traffic on camping trips. The first camping trip in which I participated, into the Anza desert, was fairly typical. We just kept driving until the car got stuck in the sand, and that determined the campsite. Most of the following day was spent in digging the car out. Such visits to the desert were (and still are) the favorite means of entertaining visitors, although some people, like Luria, will not come to Caltech unless guaranteed immunity from camping. It was on one such trip in 1950 that André Lwoff invited me to spend a year in his laboratory at the Institut Pasteur in Paris. (Benzer, 1966)

This camping trip in 1950 went down in Caltech history. André Lwoff was head of the Department of Microbial Physiology in Paris. He had spent two weeks in Pasadena working with Max:

> While at Caltech, in addition to a series of full-time experiments in the lab, the phage group saw fit to send Dr. Lwoff back to the more cultivated old world with some Wild West experiences. On a trip to the desert he and Dotty Benzer were abandoned, while the hardier element forged ahead on the ascent to San Jacinto Mountain. The two were soon hopelessly astray and in fact after some hours were beginning to despair. Mercifully some deer hunters found them and brought them back to camp. Lwoff's ho-ho's for help, which attracted the hunters, had at the same time frightened away all the deer.
>
> The high desert of Joshua Tree National Monument was chosen as the most dramatic place from which to view the full eclipse of the moon on September 25, accompanied of course by such phage trimmings as a camp fire, fried chicken and a bottle of wine. To heighten the effect Max Delbrück chose, at the phase of greatest darkness, to lead a party in a climb to a pile of rocks whose only discovered trail (directions from James Bonner) he believed he could remember and find. As the only discovered way up was soon lost, another was discovered, and the top was reached; although there were moments when, as the effects of the wine wore off, it was doubted whether the party could continue the ascent. (*Bio-Peeps*, Caltech, December 1950)

The famous Delbrück camping mania originated with Manny. In most cases, she organized the whole affair—selecting the place, purchasing the food and drinks, instructing fellow travelers—so that Max could climb into the seat of the Delbrück jeep ten minutes after he left his Caltech office and set out. Manny had become an avid camper as a student at Scripps College. Camping was, and still is, her favorite way of spending time with her family.

Such camping trips, of course, are a California specialty, with guaranteed sunny weekends and beautifully warm days all year around. These weather conditions allowed the Delbrücks to design a spacious, simple home surrounding a large garden which formed a major part of the living space. From his childhood, Max had remembered the garden of his Berlin home as a paradise. Max and Manny found a lot a block from Caltech for their California home—a home that from the very beginning was open to all visitors. Whenever possible, the Delbrücks would gather friends and colleagues in the garden for discussions or games.

The wide open spaces of Max's beloved desert and home made a strange contrast to his Caltech office (Fig. 23). He had been assigned a modest, windowless room in the basement of the Alles Building. Max claimed not to mind the atmosphere, explaining that he did not want to be distracted by a view when thinking about science.

WRONG EXPECTATIONS

Whether they approved of his behavior or not, Max fascinated the young generation of biologists and the Phage Group around him grew steadily. Two important early recruits met Max for the first

23. *Max in the California desert, at La Quinta Palm Canyon (October 1954). Photo by Manny Delbrück.*

time in 1948; in each case the personal encounter, due to faulty expectations, added to the fascination.

One was a young student from Chicago: James Watson. He was attracted to Max by the way he had revealed himself in the 1946 Harvey Lecture. When Watson read *What Is Life?* in the same year, he was spellbound. Two years later, they had a chance to meet. Watson had been a graduate student with Luria, who made the connection. From preconceived notions, Watson expected that a German professor, especially one with Max's reputation, would be bald and overweight. Instead, Max was slim and youthful in appearance and manner.

Gunther Stent had also envisioned Max as corpulent. Stent had emigrated from Berlin in the early 1930s, receiving a Ph.D. in Physical Chemistry at Chicago. After reading about Max in Schrödinger's book, he sent a letter asking if Max needed a physical chemist in phage work. Max answered (on a postcard): "We don't need someone like you."

Stent did not give up. He arranged for his own funding and contacted Max again. Since Max happened to be in Chicago, he agreed to meet Stent at Hans Gaffron's house, where Max was staying. Gaffron resembled the typical German professor. When he opened the door, Stent assumed it was Max in front of him. Immediately, Stent began to explain what he wanted to work on. It took some effort to convince Stent that Max was the lean young man behind Gaffron.

"THE PRINCIPLE OF LIMITED SLOPPINESS"

Although in the late 1940s the number of scientists working on phage and bacteria was increasing, opportunities still existed for making important discoveries by accident. The most famous example concerns a phenomenon known as photoreactivation.

The story began in Nashville. W. T. Bailey, Jr., in Max's group, had noticed that presumably dead bacterial viruses could come alive under certain conditions of infection. Bailey did not follow up on the observation, since he and others in the group failed to incorporate what Max later called "the principle of limited sloppiness."

Max coined this phrase at a 1949 meeting in Oak Ridge where this new type of revival of the dead was being discussed. Renato Dulbecco and Albert Kelner both had independently discovered photoreactivation of phages and bacteria that had been inactivated by ultraviolet light (Kelner, 1949; Dulbecco, 1949). They observed that by exposing phage particles to visible light after they had been "killed" by ultraviolet radiation, one brought them back to life. Since a laboratory is full of visible daylight—light which differs depending upon whether one is working at the window or in a dark corner—the discovery of photoreactivation explained why the plaque counts in Luria's experiments on multiplicity reactivation were often annoyingly inconsistent.

At the 1949 meeting, Max expressed his amazement that this very striking effect of visible light had not been discovered before. Many scientists had irradiated bacteria and phage and had measured survival rates. They thought they had done these experiments very carefully, under controlled conditions. But the measurements were sometimes taken during the day, sometimes during the night; at times, the plates were stacked on top of each other, and at times they were placed side by side. Sometimes there was a light bulb in a water bath; sometimes there was none. In that sense, all these experiments were conducted sloppily. But this limited sloppiness was turned into success. Kelner, working with bacteria, noticed that the lamp in the incubating bath made a difference; also Dulbecco, working with phage, noticed that the revival rate was greater on the top of a stack of plates than at the bottom.

When Max heard about this completely unexpected discovery—that light can revive life—he wrote to Luria in late 1948: "Photoreactivation is a shocker, and it is a miracle that it was not discovered before. It shows that everybody else was working too sloppily to notice it, and you . . . too precisely to encounter it. It is the old story of the principle of measured sloppiness that leads to discovery."

In introducing the symposium at Oak Ridge, Max pointed out the usefulness of the principle of limited sloppiness: "If you are too sloppy, then you never get reproducible results, and then you never can draw any conclusions; but if you are just a little sloppy, then when you see something startling you . . . nail it down."

PRIORITIES

Max did not make many friends with this suggestion, since those involved were not in a mood for jokes. They were fighting for priority. By that time, the number of people working on phage or bacteria had increased so much that competition could no longer be avoided. Until that time, the few who constituted the Phage Group formed a happy family; there was plenty of room for all, and the field was wide open. But once the crowds moved in, personal conflicts could be foreseen.

The question of proper credit had already been a slight problem for the original phage trio. In 1951, Hershey was working on a review of the phage literature; he asked Max and Luria how to give the right persons the right credit: "I am writing this letter to the two of you because together you do 90 percent of the public relations work on phage." Then he introduced "this ticklish business," for he had obviously long been irked at not having received enough credit. He wrote on May 7:

> Everybody agrees that Luria discovered multiplicity reactivation. Everybody also agrees that Delbrück and Bailey discovered genetic recombination. Hershey and Rotman followed up the latter discovery with some interesting experiments (Delbrück and Delbrück in *Scientific American* [D 38]). All these views are correct but I think they are inconsistent. Delbrück and Bailey (following up an observation of Hershey) discovered genetic recombination very roughly to the extent that they discovered multiplicity reactivation. It follows that Hershey discovered GR [genetic recombination] very roughly to the extent that Luria discovered MR [multiplicity reactivation]. Personally I think Delbrück and Luria discovered everything in 1942.

Max had to react quickly. He held the firm "conviction that the game of science is pointless if you do not put aside personal pride" (Olby, 1974). There was, of course, no problem in coming to terms with Hershey. In his answer, Max apologized first and then jokingly turned the credit question around. On May 23, he suggested the following: "Perhaps we should all pitch together and give a prize to Luria and Latarjet for *not* discovering Multiplicity Reactivation when it should have hit them in the eye from their data."

What Max referred to was an experiment devised by Luria and R. Latarjet (1947) to investigate the intracellular growth of phage. Instead of infecting bacteria with irradiated free phage particles, they applied ultraviolet (UV) light to bacteria after they were infected. After a certain amount of time, they withdrew samples and exposed them to varying doses of UV light. They did this in order to find out how many phage particles are present in a bacterium after a certain amount of time. The number of bacteria from which phages are issued can be counted; with the help of the target theory, one can calculate the number of phages inside their host cell at the time of radiation exposure.

In the case of phage T2, the fascinating result was that for a certain period, called the latent phase, there were no phage particles. The first phage particles were only detected eight minutes after infection. This was confirmed when A. H. Doermann, who had been with Max in Nashville, came up with an ingenious technique for liberating the contents of the bacterial cell prematurely. Doermann demonstrated convincingly the existence of what he called the eclipse—the time that elapses between infection and the first appearance of active particles. He showed that replication and recombination occurred during this "dark period" (Doermann, 1951). This did not clarify the true nature of the early steps in infection, but Max's early vision—that phage multiplication was "so simple a phenomenon that the answer cannot be hard to find" (D 32)—came to an end.

The eclipse proved that replication of phage was not an elementary process of a living system; instead, this reproduction occurred in distinguishable phases. What looked like a uniform phenomenon appeared upon closer examination to be a combination of events. Phage was not simple. Max could sense now that "his hunt for physical paradoxes in phage reproduction had been illusory" (Olby, 1974); in the late forties, he started to look for new priorities.

BEYOND THE GENE

By that time, Max had acquired his first graduate student. In 1949, Roderick Clayton joined him to investigate the behavior of bacteria in response to light. Max was also looking for a second such student, to do theoretical work on what he called "his little model of alternative steady states." Max had been encouraged to think about this topic by Tracy Sonneborn, who was pushing the idea that elements in the cell's cytoplasm could control hereditary traits in a genelike fashion. Sonneborn termed them "plasma genes" (Sonneborn, 1949).

As an alternative explanation, Max had conceived a steady-state model allowing gene action to be switched on and off by a system of reciprocal inhibitions among certain enzymes. When Sonneborn lectured on his plasma gene hypothesis in 1949, he referred repeatedly to Max's "entirely different mechanism" (Sonneborn, 1949). They exchanged quite a few letters in that year. In one of them, Sonneborn congratulated Max upon his election to the National Academy of Science. Max responded—on a postcard: "I never would have expected to be elected, considering what a misfit I am into any one section of the Academy. I think it's darn white of the botanists to nominate me without requiring that I pass a test in Freshman Botany!!"

VIRUSES 1950

Before Max turned completely away from phage and molecular genetics, he helped get another branch of virology on the right track. The phage work had picked up full speed because an easy plaque assay existed, in which the bacterial virus presents itself as a molecular entity, and its replication poses itself as a physical chemical problem. Practically every phage produces a plaque. For plant viruses, the efficiency is orders of magnitude lower. Simply smearing viruses on leaves and counting the lesions will produce very poor results. The situation for animal viruses was worse around 1950, though it was obvious that their analysis might have enormous medical consequences.

Around 1950 Caltech, which was devoted to basic research, was having difficulties obtaining funds for work on fundamental problems of biology. At that time a wealthy citizen of San Marino, next to Caltech, was suffering from shingles (herpes zoster). He was persuaded to offer Caltech $100,000 to start work on animal viruses. The problem of how best to use the money was given to Max. He discussed the donation with Beadle, and both decided they didn't know enough about animal viruses. To correct this, they organized a meeting of the top animal, plant, and bacterial virologists. The proceedings were published in a small volume entitled *Viruses 1950* (D 42), which Max edited.

Max felt the presentions did not describe any innovative or creative science; they offered nothing that Caltech wished to pursue. He then suggested to Beadle that they send one of the postdocs working on phage at Caltech on a tour of virology laboratories, to see if he could come up with some new ideas. Renato Dulbecco remembers that "one day Seymour Benzer and I were called to [Max's] office: Delbrück pointed out that animal virology appeared to be ready for major advances. Would either of us be interested in trying his hand at it?" (Dulbecco, 1966).

Seymour Benzer was not interested, but Dulbecco enthusiastically undertook a three-month tour. During that tour, he realized that it should be possible to make a quantitative plaque assay of some animal viruses. Although everybody else told him this couldn't and wouldn't work, Dulbecco was determined to try.

The reason for Dulbecco's certainty that a method of plaque assay could be developed for animal viruses lay in the fact that at least some of them kill cells. The problem was to get a good layer of cells, to produce a cell sheet in petri dishes that would correspond to the bacterial lawn. Eventually he worked out a procedure. He walked proudly to Max's office and invited him downstairs. Asking no questions, Max followed Dulbecco, who produced a plate with wonderful white plaques. When Max saw the holes, he asked: "What date is it? We should remember it." Somehow, though, the date was forgotten.

Although Dulbecco perfected the quantitative assay, it still was not readily accepted. He found himself experiencing the same difficulties that D'Herelle had encountered when he introduced the plaque method for phage study. He had to convince others that a plaque derives from a single virus particle, and that the linearity of the dose response means that a plaque contains a virus clone.

With the advent of this quantitative approach in animal virology, tumor virology could be advanced and vaccines could be developed. This progress was made possible by a concentration on basic research (something that Max stressed throughout his life) and by Dulbecco's willingness to move to a new field as the necessity was pointed out to him (by Max).

Max invited Dulbecco to talk about his discovery at Caltech. The seminar turned out to be a memorable event, also because of Max's involvement. Max seized the chance for a practical joke. Caltech's *Bio-Peeps* (June 1953) reported what happened:

When Biology's Renato Dulbecco agreed to give a Friday evening demonstration lecture on viruses, the C.I.T. Weekly Calendar got things mixed up a bit and announced the speaker as Ernest E. Watson, the Dean of the Faculty and Professor of Physics. This was an opportunity which

Max Delbrück could not overlook. He proceeded to make certain plans. Dr. Dulbecco had asked Dr. Delbrück to introduce him to the audience. At the proper time Dr. Delbrück arose to carry out this task. But the audience was not to hear the conventional words of introduction. Instead, Delbrück represented himself as a confused man, unable to state definitely who the speaker was to be, since the Calendar had listed Watson's name, but the subject was in Dulbecco's field. Delbrück then went on to say that, since Watson was nowhere in sight but Dulbecco was visibly at hand, Dulbecco would begin.

At this point there were small murmurs of disapproval from the audience. As Dulbecco launched into his lecture, the murmurs gathered volume and began to swell into open muttering. When Dulbecco called for the first slide, there were loud shouts of "We want Watson!" Others in the well-rehearsed claque yelled: "Who's Dulbecco?" And others: "Give us the advertised speaker!"

The lights went out and there was a sudden silence. Again Dulbecco called for the first slide. On the screen there flashed the face of Dean Watson and a voice (emanating from a tape recorder) said, "Thank you, Dr. Dulbecco for your informative introductory remarks. I hadn't expected you to be so technical or detailed. Unaccustomed as I am to speaking on viruses, I find I have very little to add to what another man said more than 3,000 years ago, 'Nothing ventured, nothing gained.' May I have the next slide?"

The voice stopped and the second slide was flashed on the screen. It was a picture of an *E. coli* bacillus. Professor Dulbecco, quickly recovering from the rush of events, said with aplomb, "And now we see a virus of more normal size."

LIFE FROM LIGHT

While Dulbecco was developing the quantitative approach that got tumor virology on its way, another important breakthrough was achieved by Max's friend in Europe, André Lwoff. In the early 1950s, he and his collaborators had found that giving a bacterium a small dose of ultraviolet light initiates the production of mature phage that eventually lyses its host (Lwoff, 1953). This phenomenon is called lysogeny. It was first discovered in *B. megaterium* and subsequently found to occur also in some strains of *E. coli*. In the case of the widely used strain K12, a phage called λ was liberated, making K12 a so-called λ-lysogenic strain of *E. coli*.

Although in lysogeny, life is initiated by light, Max reacted coolly to his friend's discovery, declaring it a "non-phenomenon" (Stent and Calendar, 1978). The reason for his resistance was the phage treaty, confining work to a common group of phages. All bacterial viruses included in the phage treaty were virulent; with them, lysogeny did not occur.

The Phage Group, confined to this limited set, did not work with phages that could enter a lysogenic cycle (temperate phages). An infecting virulent phage reproduces inside a bacterial cell and lyses it. A temperate phage can either do this or insert its genes into the bacterial chromosome. In the latter case, it becomes what is called a prophage.

The importance of this insertion is very obvious today. When a prophage gets back on the road and starts to multiply in the bacterium, it can take some bacterial genes along. The progeny phage then carries bacterial genes. If this new phage infects another bacterium, it has transferred bacterial genes from one bacterium to another. This process, called transduction, forms the basis of gene technology.

Max was not interested in the technical possibilities of transduction; in his view, a special explanation did not seem to be required, "for what is transduction but a sort of sexual communication between bacteria mediated by sperm-shaped phage particles?" (I) Max turned to lysogeny only in order to see if this effect would allow one to distinguish between any of the mechanisms that had been suggested to explain mutual exclusion and interference. Some experiments that he developed with Jean Weigle in 1951 did not resolve this question; all these did was to "furnish conclusive proof that mutual exclusion must involve some mechanism other than the establishment of a barrier to penetration" (D 44).

Since Max's pursuit of a clear paradox in genetics led to increasingly complicated situations, he turned his attention to other model systems. He left genetics to start on a new venture, just a few weeks before molecular biology reached its triumphant peak.

The Great Denouement

A WOMAN IN BIOLOGY

Biology was revolutionized in the 1950s with the help of bacteria and their viruses. The shift of attention was noted in Milislav Demerec's introduction to the 1951 Cold Spring Harbor Symposium dedicated to "Genes and Mutations":

> One of the most remarkable developments of these ten years concerns the organisms used in gene studies. In 1941 about thirty percent of the Symposium papers reported research carried on with *Drosophila,* and only six percent dealt with microorganisms; whereas this year only nine percent of the papers relate to *Drosophila,* and about seventy percent to microorganisms.

Ten years before, Max and Luria had begun their collaboration; the Phage Group they initiated dominated the early stages of molecular genetics. As a result, the way genetics was done had changed considerably. This transformation would eventually unravel the physical properties of genes, but when biologists gathered at Cold Spring Harbor in 1951, "the original problem of defining the unit of heredity, which almost fifty years ago was designated 'the gene,' [had] not yet been solved" (Demerec, 1951).

The new way of thinking about genetics that the Phage Group had induced distracted attention from the seminal work of Barbara McClintock, who reported her observation of movable genetic elements in 1951. "Her talk [in Cold Spring Harbor] was met with stony silence" (Fox-Keller, 1983), since she spoke about the rearrangements of entities whose nature was still an unresolved problem. An additional barrier arose from differences in approach and focus. Whereas McClintock was attempting to explain an organism, the molecular biologists were seeking to understand their presumed genetic molecules.

The complementary approaches to scientific questions are further reflected in the persons of Barbara McClintock and Max. It's not only that they were so different in their outer appearance—she was a small woman, he a tall man—but that their ways of being scientists were so different. Whereas Barbara McClintock worked in solitude, developing a "capacity to be alone," Max needed a group to talk to.

Max undertook quite a number of different research programs, all directed by his desire to find an organism and a biological phenomenon that were simple enough to suit a physicist but able to reveal fundamental questions in biology. He thus concentrated on an idea, selecting his experiments accordingly. For him, the concept came first.

On the contrary, Barbara McClintock began to work with corn and stayed with it. She was a naturalist. In her view, the material came first: "Models . . . are so bothersome" (Fox-Keller, 1983). Max was a physicist who loved a good theory as much as an elegant experiment. He opted for good ideas even in the sense that he preferred a clever mistake to a dull truth. Max was by far the more influential of the two.

A FAMILY AFFAIR

A large part of the explanation of Max's influence resides in his style of leadership. By integrating collaborators into his family, Max created not just a Phage Group, but a Phage Family. In this family, Max was a father with the playfulness of a child.

In their Pasadena home, the Delbrücks lived isolated from American commercial culture; they had no television, no radio, and no magazines. They never owned a large American car—long before small cars became popular, the Delbrücks drove functional, modest automobiles. Their furniture was always simple but elegant. The living room was dominated by a grand piano, some Indian rugs, and Jeanne Mammen's paintings.

Max's first son, Jonathan, was born in Tennessee; the first daughter, Nicola, was born in Pasadena. The birth of the second child was the reason Max missed the 1948 summer in Cold Spring Harbor. To compensate, he arranged an extra phage meeting in the West that year. Later, in the early sixties, the Delbrück family expanded further. A second son, Tobias, was born in Pasadena, and Manny Delbrück gave birth to their fourth and last child, Ludina, in Cologne, West Germany. Max so enjoyed this two-installment family that he recommended to a number of colleagues and friends that they do likewise.

Within the Delbrück family, Max definitely was not the leader, for he took orders from his wife. Manny ran the family, with one exception: Max insisted that the Delbrücks should never own a dishwasher. He felt that dish-cleaning chores provided a good opportunity for talking to each other in a relaxed, informal atmosphere; he enjoyed dish-cleaning time as a family affair.

The essence of Max's idea of a family was the communication within it. Max would never talk about his family, but used every opportunity to get the members to talk to each other. This led to some unconventional customs. One most easily sees this in the way the Delbrücks celebrated Christmas. Every family member was supposed to have presents for the others, but these were preferably not purchased items. One parcel was given at a time in front of the others, and the recipient had to guess what the parcel contained before opening it. He or she was given clues and, depending on the number of clues required before the right answer was found, the recipient was classified on a scale ranging between "genius" and "world's greatest idiot." The clues hung on the tree, serving as decoration. The wrapping paper was always saved for reuse at the next celebration; the Delbrücks extended this careful handling to all items of daily use.

Family retreats, such as camping trips, generally included colleagues and friends. On camping trips, Max used the long hikes to talk to those he had not had time to speak with in the laboratory. In these undisturbed conversations, Max not only talked science but also inquired about families. His opening question, if he had not talked to a person for some times, was frequently, "What are the scandals?" Max loved gossip, and would contribute whatever he knew.

Camping trips provided an opportunity for Max to play numerous practical jokes. His favorite was perpetrated at the so-called Double Surprise Canyon, in the Mohave Desert near Palm Springs. Max would lead a group into a canyon which seemed to be a dead end (the first surprise). Unbeknown to the guests, Max not only was equipped with a map but also had taken this route several times. After some pretended deliberation with his wife, he would decide that they would not turn back but climb a rock that was in the way. This was not a simple thing to do, and quite a few guests ended up with torn shirts and scratches. Finally, Max also climbed up, only to declare that now there really was a dead end. From the elevated position, one could see a rather large cleft at one side of the rock. Asking, "Which way do you think the water will follow?", Max disappeared under the boulder. This was the second surprise, for indeed a dark, narrow canyon led finally into a wide open wash, on which one could cross the desert back to the campsite.

On these camping trips, during clear desert nights, Max never tired of explaining the stars. He actually took some measurements, without complicated optical devices, in order to catch up with the ancient astronomers and to get a feeling for the complexity of the movement of the stars, as the ancients must have done.

On each trip lasting more than a day, the Delbrücks kept a diary. Each day, a different Delbrück was selected to summarize the activities and the adventures. Some of Max's entries—after listing who had been doing the chores—dealt with his astronomical interest. In 1971, he noted:

> Up at dawn, scrambling up hillside to get clear view to N.E. to find Venus at 5:40, and wait for sun to pass some spot, 60′ later. Finally resolved mystery of slow motion of Venus, relative to sun, during last week: *not* because of error of calculation of expected speed, *not* because I had mistaken Mercury for Venus, *but* I had mistaken by thought, stupidly, that Venus was approaching lower conjunction, rather than upper one. Somewhat of an anticlimax.

Sometimes the nights got very cold, but Max was always able to devise unconventional solutions: "I eventually relieved the discomfort of freezing in my sleeping bag by inventing the SWEATER TROUSERS, with my legs in the arms of my sweater."

"ALL DNA AND NO PROTEIN . . ."

Due to the success of the Cold Spring Harbor summer courses and of Max's recruitment efforts at Caltech, the Phage Group grew. In the early fifties, it seemed time for an international conference. André Lwoff organized one in France. In 1952, at least fifty members of the Phage Group gathered at the Abbaye de Royaumont near Paris for the first International Phage Symposium. The great discovery reported there was the result of an experiment by Alfred Hershey and Martha Chase. In their so-called blender experiment, they had demonstrated that the DNA of phage contained *all* genetic information. This went beyond Avery's result of 1944, which showed that DNA contained *some* genetic information.

Only a few people were primarily interested in phage DNA in the early 1950s. The admissible view of the time was presented by Luria (1950). A phage particle was considered a dual structure (mainly protein and DNA). After phage infection, the bacterial metabolic system falls under the control of the nuclear apparatus of the phage, producing mainly phage-specific materials. The infecting phage particle does not itself survive the infection, losing at least its ability to infect another bacterium.

A year later, the thought slowly occurred to phage workers that the nucleic acids may form the essential part of the phage particle and that the protein portion may be necessary only to allow entrance to the host cell.

In a letter dated November 16, 1951, Roger Herriott wrote to Hershey: "I've been thinking—and perhaps you have, too—that the virus may act like a little hypodermic needle full of transforming principles; that the virus as such never enters the cell; that only the tail contacts the host and perhaps enzymatically cuts a small hole through the outer membrane and then the nucleic acid of the virus bead flows into the cell" (Hershey, 1966).

Hershey proved this idea to be correct in the blender experiment (Hershey and Chase, 1952) which was reported at the Abbaye Royaumont meeting. In his summary Max pointed out that with this discovery, "we are thus confronted with the paradox that phage enters its . . . cycle containing all DNA and no protein, and seems to emerge from the cycle all protein and no DNA."

So, at the end of the meeting in France, phage workers knew that phage DNA is the sole carrier of the hereditary continuity of the virus. The details uncovered from this point on concerning the physiology and genetics of phage reproduction would have to be understood in terms of the structure and function of DNA.

THE DOUBLE HELIX

In the summer of 1952, one of the participants of the first International Phage Symposium was already working hard to elucidate the structure of DNA, which he never doubted as being the genetic material. James D. Watson had been polarized toward finding out the secret of the gene from the moment he read *What Is Life?* He became a graduate student of Luria's in Indiana, where he shared an office with Renato Dulbecco. Here Watson's research project involved determining whether phages that had been inactivated by X-rays gave any multiplicity reactivation.

This was in 1948, when the secret of the gene was not even close to being uncovered; the arguments circled around which approach would provide insights. Within the Phage Group, there were two opposing opinions on how to proceed. Seymour Cohen, who had taken the second phage course given by Max in 1946, advocated turning primarily to biochemistry to explain genes, while Luria and Max opted for concentrating on a combination of genetics and physics. Watson had to make a choice. When he was about to finish his thesis on X-ray inactivated phages, he showed his data to Max. Watson remembered Max's reaction of 1949:

> Delbrück, like everyone else, was only mildly interested in my results but told me that I was lucky that I had not found anything as exciting as Dulbecco had, thereby being trapped into a rat race where people wanted you to solve everything immediately. If that had happened, he felt I would lose in the long term by not having the time either to think or to learn what other people were doing. (Watson, 1966, in PATOOMB)

Jim Watson felt that if one wanted to understand genes, it would not hurt to learn something about the substance they were made of. As a consequence, he planned to spend his postdoctoral time in a biochemically oriented laboratory in Europe. Luria suggested that he work in Copenhagen with Herman Kalckar, who had taken the first phage course back in 1945 (Fig. 24).

Watson thus departed from the mainstream of the Phage Group, not in spite of his tremendous admiration for Max but because of it. He felt he "could never do anything important" if he tried to think along the same lines as Max did with "that high powered brain" (Watson, 1966, in PATCOOMB).

24. Max is welcomed at Copenhagen Harbor as he arrives for the Polio Congress in September 1951 (The sign reads "Welcome Max, ...) Left to right: Gunther Stent, Ole Maaloe, Max, C. Bresdi, and James Watson. Watson was about to end his postdoctoral stay in Copenhagen, where he worked with Herman Kalckar on biochemical approaches in studying phage. Just after the Congress, Watson switched to Cambridge, where he began using structural chemistry and X-ray crystallography to develop a model of the structure of DNA. Photo courtesy of Gunther Stent.

After spending some time in Copenhagen, Watson realized that he had erred in selecting biochemistry as a source of information; he transferred to Cambridge, where he could learn and use both structural chemistry and X-ray crystallography. Before Watson left

Copenhagen for England, Max offered him a research position in the Biology Division at Caltech with funding from the Polio Foundation; Watson was to start in September 1952. When Watson wrote Max in the summer of 1952 that he wanted another year in Cambridge, Max responded in a typical dualistic fashion, as described by Watson: "Without hesitation [Max] saw to it that my forthcoming fellowship was transferred to the Cavendish," but he "threw up his hands in disgust" that Watson was doing the wrong thing (Watson, 1968). Watson understood Max's indifferent response to his DNA model building, since "in Delbrück's world no chemical thought matched the power of a genetic cross" (Watson, 1968).

From Cambridge, Watson kept Max informed by writing at least one long letter a month to Pasadena. These letters both presented Watson's ideas and kept Max up to date with the gossip. Max became particularly interested in the structure of DNA because it would have helped the theory he was developing. Max was working on what turned out to be his final contribution to phage genetics. He was trying to understand the recombination frequencies that had been obtained under various experimental conditions. In order to do this, Max applied the methods of population genetics by assuming that phage genomes exist in an intracellular mating pool. Such a theory had to be a losing cause as long as the structure of the genetic material still eluded scientists. But with Watson's move to Cambridge, progress in this direction came rapidly.

By late 1951, Watson was already very optimistic about a solution for the structure of DNA. He wrote to Max on December 9, "we believe the structure of DNA may crack very soon. . . . Our method is to completely ignore the X-ray evidence." By "we," Watson meant himself and Francis Crick, whom he described in the same letter as "the most interesting member of the group": "[Crick] is no doubt the brightest person I have ever worked with and the nearest approach to Pauling I've ever seen—in fact he looks considerably like Pauling. He never stops talking or thinking, and since I spend much of my spare time in his house . . . I find myself in a state of suspended stimulation."

Watson always felt that Linus Pauling would be the one to beat in the race for the DNA structure; he hoped that Max would keep him informed about the ideas Pauling developed at Caltech. Pauling had become famous for elucidating the α-helix as a secondary structure for proteins (Pauling and Corey, 1952). He conceived of a DNA structure that he submitted for publication on December 31, 1952; it appeared in the February 15, 1953, issue of the *Proceedings of the National Academy of Sciences* (Pauling and Corey, 1953). When Pauling described his model in a seminar on March 4, 1953, at Caltech, he could not convince Max. Indeed, Max had been informed by Watson on February 20:

> I am now extremely busy, largely working on DNA structure. I believe we are close to the solution. We have seen Pauling's paper on nucleic acid. Have you? It contains several very bad mistakes. In addition we suspect he has chosen the wrong type of model. All in all, however, Pauling's paper is at least in the proper mood and the type of approach which the people of Kings College London should be taking instead of being pure crystallographers. I had started on DNA when I first arrived in Cambridge but had stopped because the Kings group did not like competition or cooperation. However, since Pauling is now working on it, I believe the field is open to anybody. I thus intend to work on it until the solution is out. Today I am very optimistic since I believe I have a very pretty model, which is so pretty I am surprised no one has thought of it before. When I have the proper coordinates worked out, I shall send a note to Nature, since it accounts for the X-ray data, and even if wrong is a marked improvement on the Pauling model.

The model mentioned but not described in the letter presented DNA as "a two-chain affair." Watson had envisaged that each DNA molecule consisted of two chains with identical base sequences held together by hydrogen bonds between pairs of identical bases. After the seminar, Max could tell Pauling that Watson and Crick had new ideas for the DNA structure, but he didn't know what the ideas were. Pauling was very interested and asked Max to keep him informed, which Max promised to do.

By March 1953, the solution conceived by Watson was ready. Watson described his model of DNA's structure in a long letter to Max dated March 12. DNA was found to be a double helix. Two sugar-phosphate backbones twist about on the outside, with the flat hydrogen-bonded and stacked base pairs forming the core. The structure of DNA resembles a spiral staircase, with the base pairs forming the steps.

At the end of the letter, Watson asked Max not to mention anything about the structure to Pauling. He and Crick would do this themselves, by mailing Pauling a reprint. Max now faced a dilemma, since he had promised Pauling to tell him what structure Watson and Crick proposed the minute he heard. Max did call Pauling and showed him the letter; Pauling was convinced within five minutes.

THE DENOUEMENT

The discovery of the structure of DNA by Watson and Crick and of the proposed mechanism of its replication provided phage workers with a party line, as Max used to call it. That is, major consensus was reached concerning the life cycle of phage and the central genetic role of DNA; biologists now had a chemical structure for DNA that said something about function. However, it did not provide Max with an opportunity to reveal the epistemological lesson in biology. The complementarity of the base pairs of DNA should not be confused with the deep complementarity that Niels Bohr had predicted and that Max was looking for.

When Max saw the double helix explanation, he felt as if he had been struggling with a chess problem only to be shown the embarrassingly simple solution. Max felt foolish when he realized that the double helix revealed "with one blow the mystery of gene replication" (D 95). Nevertheless, the structure was marvelous—Max himself was fascinated by its beauty—and it made the whole business of replication look like "a child's toy that you could buy at the dime store." Everything "was built in this wonderful way that you can explain in *Life Magazine,* that really a five-year-old can understand what's going on—that there was so simple a trick behind it. That was the greatest surprise for everybody," as Max told Horace Judson in an interview in 1972. It was the great denouement.

The publication of the structure of DNA will rank in science history with Newton's theory of gravitation or Darwin's suggestion of an evolutionary origin for species. Unlike other triumphs of scientific imagination, it had an immediate effect, since it provided the answers to questions on many people's minds. There was no confusion like the one that complicated the success of quantum mechanics, and the theory was not forbiddingly difficult as was Einstein's theory of relativity. DNA was simple and the early molecular biology that emerged from its discovery repeatedly confirmed "simple expectations of nature" (I).

The unraveling of the DNA structure cast light on the mechanisms of hereditary by suggesting a biochemical mechanism for replication. It made obvious that there must be a genetic code, and it allowed one to predict messengers that carry information. Filling in the details produced molecular biology, which has led in turn to the deliberate manipulation of genes. Thus, the double helix is to date the most important and influential discovery in biology. It will probably make its discoverers famous as long as there are biologists.

THINKING ABOUT REPLICATION

Shortly after Max learned about the double helix, he informed Bohr about it, comparing the stature of this breakthrough with that of Rutherford's discovery of an atomic nucleus. Max also wrote to Watson on the same day (April 14, 1953):

> I have a feeling that if your structure is true, and if its suggestions concerning the nature of replication have any validity at all, then all hell will break loose, and theoretical biology will enter into a most tumultuous phase. Only part of this will involve chemistry, analytical and structural. The more important part will consist of attempts to take a fresh view at the many problems of genetics and cytology which came to dead ends during the last 40 years.

Max himself concentrated on replication, his old favorite. In a review that he wrote in collaboration with Gunther Stent, the term "semi-conservative replication" made its first appearance (D 49). It denoted the mechanism that Watson and Crick had suggested. Such a process would require a complicated unwinding of the DNA thread; Max felt this suggestion would encounter serious dynamic difficulties. To overcome these difficulties, he proposed a mechanism in which the nucleotide strands would first be cut and then repaired (D 47). Such a break-and-reunite process was actually found to exist many years later (Kornberg, 1974). When M. Meselson and F. Stahl (1958) designed an experiment that proved semi-conservative replication to be the actual mechanism of DNA synthesis, Max characteristically didn't believe a word; he made them run the ultracentrifuge twenty-four hours a day until they could wipe out any doubts.

At that time, none of the enzymes involved in DNA replication was known. In recent years, a considerable number of enzymes have been found that perform cutting and gluing. These topoisomerases, gyrases, and other enzymes work differently from what Max proposed in his model of 1954. Today, DNA synthesis is by no means a resolved problem. The debate still continues as to how the intertwining of the strands is accomplished; on the whole, though, the most basic feature—that the replication is done by synthesis of complementary strands—is seen as very clearly true. Nevertheless, even two years before the thirtieth anniversary of the discovery of the double helix, the mechanism of DNA replication was not known in sufficient detail to satisfy Max.

EARLY THINKING ABOUT RECOMBINATION

Max's last resort for seeking complementarity in genetics was the mechanism of recombination. Even now this process remains quite a challenge. In the early 1950s, Max tried to work out a theory of recombination events in phage based on multifactor crosses. As N. Visconti remembers, Max developed this scheme mostly in front of blackboards:

> He could stand for hours in front of one, writing down and then erasing complicated matrices on which he worked before coming down to simpler methods. When he had settled a step in the theory, he would copy on paper the result of the calculations, erase the blackboards and start on a new step. People would walk in and out, ask for an explanation, or make a suggestion. Delbrück was rarely disturbed by intruders, and could easily talk about what he was doing while doing it. He was also not disturbed by suggestions, even if they were wrong, as long as they followed some logical pattern. In a sense his teaching was Socratic. (Visconti, 1966, in PATOOMB)

In January 1952, Visconti joined Max in this effort; together they published "The Mechanism of Genetic Recombination in Phage" (D 45). Their theory placed the genetics of phage very much in line with the classical Mendelian theory which had been worked out for higher organisms. However, for phage, one cannot study the recombination directly from a single mating. By applying the methods of population genetics, Max and Visconti could calculate the recombination values for a single mating and thus build linkage maps.

Max was hoping that new insight into biological mechanisms would emerge from the theory and that analyzing recombination could provide clues as to the nature of the gene and the limitations in its understanding. This hope never materialized.

Max never published anything else on phage genetics; for the rest of his life, he stayed away from the science whose methods he had been using—population genetics. After their joint work, Visconti wanted to quit science altogether. He explained to Max that he felt inferior to all the brilliant scientists working at Cold Spring Harbor. Max's reply was pungent: "You don't have the inspiration or the talent to be an artist. Then what else do you want to do in life besides be a scientist?" That was argument enough.

LATE THINKING ABOUT RECOMBINATION

In the late 1950s, Max was still stating publicly that complementarity in biology "might just be around the corner"; the event of recombination seemed the best bet to encounter such a situation. In 1957, he wrote to Bohr: "The more I think about the possible relevance of the complementarity argument for biology, the more I seem to be driven to the conclusion that it can be fruitful only if it comes in some such manner as I tried to illustrate in the example on genetic mapping."

Max here was referring to a lecture he had given at the Massachusetts Institute of Technology. In November 1957, Niels Bohr had delivered six Karl Taylor Compton Lectures at MIT on "Complementarity in Quantum Physics." Max was invited to present a seminar in the life sciences on the day after Bohr's lecture, on complementarity in relation to biology. Max made an effort to discuss this in concrete terms, picking recombination as his topic. This process allows scientists to draw what is called a genetic map: one can deduce the arrangement of mutable sites (genes) on a chromosome from recombination experiments. The issue Max discussed in 1957 was whether the genetic map is a direct image of the sequence in the genetic molecule (DNA). In this situation, one attempts to correlate physical parameters—the sequence of nucleotides—with nonphysical parameters—the genetic map distances.

Genetic mapping is done by examining mutations. A mutation marks a point on the genetic map; distances between these points are measured in terms of recombination frequencies drawn from

experimental crosses between pairs of organisms carrying prede-
termined genetic markers. Of three hypothetical markers, A, B, and
C, the two which give the greatest recombination frequency are put
on the outside, thus supporting a statistical conclusion that they
represent separate genetic units that can recombine during crossing over.

Molecular mapping on the other hand is the nucleotide sequence
analysis of purified DNA by chemical methods. It consists of observations made under experimental conditions which exclude any
recombination event. Recombination occurs only in intact cells
and not in those prepared for sequence analysis. The conditions
under which recombination takes place are incompatible with conditions that define the physical state. In the 1957 lecture, Max made
clear his hope that the genetic and the physical map would prove
complementary to each other; sequence information was not even
on the horizon at this time.

Nevertheless, the question did not remain open for long. The
agreement of a genetic map with the actual distances along a DNA
molecule was shown when it became possible in the early 1960s to
locate the sites of specific mutations along a DNA molecule with the
help of electron microscopes (Watson, 1975). With genetic and
physical maps found identical, genetics did not open the door to
complementarity.

EXPLOITING RECOMBINATION

At the time Max was thinking about recombination, another of his
disciples, Seymour Benzer, was using this phenomenon to study
the genetic fine structure of one particular locus of the phage
genome. Benzer's studies led him to reform the classical concept of
the gene. The classical view presented an indivisible gene defined
by the one-gene-one-enzyme relation. Benzer created the new "molecular" gene, which finally bridged the conceptual gap between
geneticists and biochemists.

Benzer had been looking for "high-resolution" genetic mapping
in order to see if various existing definitions of the gene were
equivalent. What he needed was a system in which he could detect
a small number of recombinants. This became possible by way of
the so-called rII phenomenon that had been discovered by Hershey.
It provided a method to "run the genetic map into the ground," in
Max's words. With his procedure, Benzer could resolve or find

mutations that were located at adjacent nucleotides. For this break-through, he had to depart from the Phage Group's party line, for he had to take conditionally lethal phage mutants seriously.

With the help of these rII mutations (two classes of T2 mutants), Benzer succeeded in providing an experimental definition of the gene. He introduced the term "cistron" as a genetic unit of function. As Benzer described his results:

> These rII mutants all mapped within a few percent recombination of each other, but were located at a large number of distinct sites. A unique order of the sites could be established, and the map could be sharply divided into two contiguous segments or *cistrons,* as I later called them, such that any mutant located within one cistron would functionally complement any located within the other cistron. (Benzer, 1966, in PA-TOOMB)

When Benzer showed Max the data in the summer of 1954 in Amsterdam, Max's comment was, "Delusions of grandeur." But he submitted Benzer's paper to the *Proceedings of the National Academy of Science,* predicting this problem would keep Benzer busy for the next ten years.

Mapping in the rII region was what members of the Phage Group called "Hershey Heaven." When asked for his idea of scientific happiness, Hershey is reported to have answered, "To have one experiment that works, and do it every day for the first time." The Hershey-Chase experiment was Hershey's experiment that worked. Benzer for a time enjoyed running the genetic map into the ground with the rII region. He retreated from this heaven after Max appended a footnote to a letter that Manny Delbrück had written to Benzer's wife:

> Dear Dotty,
> Please tell Seymour to stop writing so many papers. If I gave them the attention his papers *used* to deserve, they would take all my time. If he *must* continue, tell him to do what Ernst Mayr asked his mother to do in her long daily letters, namely, *underline what is important.*

THE CODE GAME

In the mid-1950s, the DNA structure posed another fascinating problem. How is the information that is stored in the DNA used to make proteins? This question was resolved in the 1960s, to the extent that we now know how an amino acid sequence in a protein is encoded in a nucleotide sequence of DNA.

Max remembered the first attempts to understand the making of proteins:

> how this information (that is stored in the DNA) is really used to code genetic information[,] that was totally nebulous in the beginning; the first incredibly bold attempt to cut through this fog was George Gamow's [1954] who suggested a direct method of using DNA as a template; to lay the amino acid sequence on it in the big groove of the DNA. That was manifestly wrong in its chemical details, and was therefore rejected by everybody including me, but surprisingly in the end it turned out to be relatively close to the truth. Of course, that was before transfer RNA and ribosomes and all this other jazz had been discovered. But the general principle that the coding follows such simple rules turned out to be true, and that was a very great surprise to me. I was certainly not prepared to believe in such a simple procedure. (I)

In his 1954 paper, Gamow pointed out that nucleic acids could be viewed as a sequence of four different bases or nucleotides. The physicist Gamow found it obvious that the four nucleotides were the symbols of a code, the genetic code. Nine years before, another physicist, Schrödinger, had inconspicuously introduced a crucial notion: he called the chromosomes "the hereditary codescript," postulating that these visible structures would "contain in some kind of codescript the entire pattern of the individual's future development and of its functioning in the mature state" (Schrödinger, 1945).

The question to be solved in 1954 was which sequence of bases in DNA would code for an amino acid residue in a protein. Since living cells used a total of twenty amino acids to build their protein, Gamow realized that each code word had to be a sequence of three bases, called a triplet. With four symbols, one for each nucleotide, only sixteen possible duplets can be constructed; but there are sixty-four combinations of triplets. Gamow concluded that the genetic coding occurs in triplets and further that there are several code words for each amino acid, i.e., that the code was degenerate. Both conclusions have been found to be true.

Before such consensus was achieved, biologists tried to construct a nondegenerate code with a unique triplet coding for each amino acid (Crick, Griffith, and Orgel, 1957). In order to cut the number of meaningful triplets from sixty-four down to the magical twenty, Crick and his coworkers assumed that there was only one way to read a sequence, that one had to start at a particular DNA residue and not one or two bases further out. Only certain triplets were assumed to possess a meaning and, together, to constitute a "dictionary" of "words." Their sequence would then form the message. All messages generated by reading this sequence from the incorrect starting point, from within a triplet, for instance, would be meaningless, i.e., these words would not exist in the dictionary. Such a code was termed "comma-free."

It could be shown that in the case of four letters and words of length three, the largest comma-free dictionary has exactly twenty words, the magic number of amino acids (Crick, Griffith, and Orgel, 1957). This paper supported the notion that the genetic code was comma-free.

Max had asked two mathematicians about the problem of finding other possible comma-free codes, where triplets do not have to be separated from each other by commas. In a paper entitled "Construction and Properties of Comma-Free Codes (D 51), S. W. Golomb, L. R. Welch, and Max described procedures for constructing all comma-free triplet codes involving the maximum number (twenty) of triplets.

The paper attacked quite a few general problems. Unfortunately, it was doomed. The very first series of experiments to determine the occurrence of repetitions of symbols in nucleic acids showed that the same nucleotide did occur four and even five times in a row. Since one nontrivial result of the mathematics of comma-free codes (D 51) was that in any message constructed from any of these dictionaries, the same nucleotide symbol could not occur as many as four times in a row, the comma-free triplet thesis was gravely wounded, never to recover.

But there is an interesting and amusing postscript to Max's love affair with the genetic code, as described by S. W. Golomb (1982) and confirmed and detailed by W. Shropshire (in a letter to S. W. Golomb from September 1982). In the fall of 1958, the Nobel Prize for Medicine and Physiology was awarded to George Beadle, who at the time was in England, on leave from Caltech. The day of the radio announcement, Max showed up at the laboratory with a comma-free code message that he sent as a cable to Beadle. Max passed out copies of the message to those working in the laboratory, regaling them with stories of his difficulties in convincing the cablegram operators that he was not a Communist spy sending out high-level secrets. The message read as follows:

> To: Beadle
>   Botany
>   Oxford, England
>   ADBACBBDBADACDCBBABCBCDACDBBCABBAADCACABDABDBBB
>   AACAACBBBABDCCDBCCBBDBBBAADBADAADCCDCBBADDCACAA
>   DBBDBDDABBACCAACBCDBABABDBBBADDABDBBDABDBACBAD
>   BBDBACBBDCBBABDCACABBACDAACADDBBDBBBADDBADAxBBA
>   DDBADAACBCDCACABBABCABCBBDBACBDDACDBDDCBDC

The conspicuous x in the third line of the text enabled Max's collaborators to decipher it within thirty minutes. It read:

> Break - this - code - or - give - back - Nobel - Prize - Lederberg - go - home - Max - Marko - Sterling

The message was sent by Max, Marko Zalokar, and Sterling Emerson from Caltech. Joshua Lederberg also received his Nobel Prize in 1958, which explains the reference to him.

Max based the telegram to Beadle on a comma-free code consisting of twenty triplets from the four-symbol alphabet A, B, C, D. He did the encoding first by looking at all the distinct letters in his proposed message, counting "space" and "X" and "Z" as letters. Since there were twenty-one symbols, and only twenty code words, he decided to send the "x" of Max" unencoded.

The next day, a message arrived in response. Max rushed into the laboratory, passing out numerous copies. He offered a prize of five dollars to the first person who could decipher the text. Shropshire brought it home to his wife, who cracked the code simply by alphabetizing the three letter sets and assigning plain letters (this code was not comma-free).

> To: Max Delbruck
>   Biology Division
>   California Institute of Technology
>   Pasadena, California
>   G W B T O M D I M S U R E I T S A F I N E M
>   BBADDCACADDACCBBDCADABCBBDCCDCDDBCDBADBBCBDDAC-
>   DCABAADCBCBCCAADBBDC

```
E S S A G E I F I C O U L D D O T H E F I N
ADBCDCCDCABABBAADBBCBADCBCBACBCCBDDBBDBADAADAC-
CBDDABCAADBADCBCBCBCCAABAB
A L S T E P
DBCDCDDAADBCDA
```

There is one addendum to the coding game at Caltech. At the Nobel ceremony for Beadle, a special double helix made of toothpicks in four colors was presented to the laureate. Decoded from its triplet code, the message read: "I am the riddle of life know me and you will know yourself."

MORE CAUSE FOR CELEBRATION

The genetic code was discovered in the 1960s, more than a decade after Max been active in phage work. But he was by no means forgotten by his former fellow geneticists.

In 1966, at Max's sixtieth birthday, the former members of the Phage Group prepared the Festschrift in his honor entitled *Phage and the Origins of Molecular Biology,* which Max referred to as PATOOMB. The book was supposed to be handed over to him at Cold Spring Harbor, where Max was spending his summer working. When he refused to attend the presentation ceremony, Gordon Sato was delegated to persuade him to come. Sato remembers the occasion:

> I found Max out in the dark behind Blackford Hall. It appeared that this whole occasion had somehow driven him into a state of depression and that he was pondering the significance of his life and, I guess, the life of Everyman. He knew instantly why I had come. When I came up to him, he said, "No single man makes any difference." So I pursued the only argument I could and said, "Look, Max, this Festschrift isn't for you; it's for your friends, and you have to attend." And so, although it was against his principles, that time he managed to bend.

Max attended the meeting.

Gordon Sato, now a professor of biology in San Diego, is Max's most successful graduate student. Back in 1950, he was a Japanese gardener in Pasadena, mowing lawns in the vicinity of Caltech. He wanted to become a student, but when he tried to enroll he encountered difficulties. Eventually he ended up in Max's office, recounting the story of his life. Max realized that this was an outstandingly intelligent and dedicated young man who needed a strong backer because of his poor preparation and lack of funds. Max assumed the responsibility; he managed to bend a few rules to take Gordon Sato on as a graduate student, deciding to accept "somebody who shouldn't have been there" (Sato). He had to support Sato for years because his grades were too poor in the beginning to get outside financial support. In a letter to T. F. Anderson (in November 1954), Max explained why he had helped Sato: "My weakness towards him has always been motivated by the idea that during my impecunious student days I was just as disorganized and uneven in my perform-ance, and that thereafter I had no right to cut his career short."

Later Sato, among others, discovered that one can grow different cell lines in serum-free media under controlled conditions. They require different mixtures of hormones and growth factors (Haya-shi, et al., 1978). Sato was involved in establishing many important differentiated cell lines.

"THE HONEYMOON IS OVER"

The Festschrift and its many reviews began the discussion of the history of the new genetics. In the following year, one of the editors of PATOOMB, Gunther Stent, composed a review of the reviews. In it, he described his view of the "Molecular Biology That Was," proposing discrete stages in the development of the field. In partic-ular, after a "romantic" beginning and a "dogmatic" period, the whole enterprise now was in a boring "academic" phase.

In the meantime, however, starting in the mid-seventies, new revolutionary features of the gene have been discovered through technical breakthroughs. These in turn have allowed the rise of gene technology. Molecular-biology has now entered a risky stage and, in Max's judgment, "the honeymoon is over" (I).

In the same year that Stent published his systematic view of the golden age of molecular biology, a very personal account appeared in print, becoming an international bestseller. In 1968, Jim Wat-son's description of the discovery of the structure of DNA, *The Double Helix,* was published. Because Max influenced Watson so much, Max's reaction to the book is of interest, especially since the book received heavy criticism from many scientists. It was attacked for the harsh things it had to say about individual scientists, and also because it suggested that science was more a race for personal honor than a search for knowledge. Scientists on the whole held very firm opinions, pro or con, as to the portrayal of scientific activity given in *The Double Helix.*

Max felt that "Jim's gossip book" was an important confession on Watson's part, and that he had written more damaging things about himself than about anyone else. Max had never been concerned with receiving acclaim for his discoveries, but he did feel the presentation of scientific motives was accurate for certain scientists, though not for all. According to Max, "the way the development of science is represented in most textbooks is completely asinine." Science does not progress from hypothesis via experiment to conclusion; instead, the progress of science is tremendously disorderly, and the motivations leading to this progress vary tremendously. So do the reasons governing entrance to science.

The Restless World of a Scientist

"It is very satisfactory to work entirely by yourself for a while with not a person in the world sharing your interest." (Courtesy California Institute of Technology, Pasadena, California USA)

Looking for Biophysics

BEYOND THE GENE

Max's admonition—"A good scientist has to tame his curiosity" (I)—describes his own attitude toward conducting research. In genetics, for instance, success meant concentrating on a limited set of phages and analyzing a small number of phenomena. With respect to the whole of science, however, Max did not restrain his curiosity. He gave attention to a wide range of scientific developments, following them with the joyful optimism of one who looks forward to new insights.

Another reason for indulging in a variety of scientific domains was the "ulterior motive" that made Max turn from physics to biology. Max wanted to apply the lesson that atoms provided to the problem of life. Since in the 1930s the best potential lay in genetics, Max started there, but it was only one of many possibilities.

When Max returned to Caltech in 1947, he went there to stay. But he was only permanently settled in a physical sense; at Caltech, Max continued to pursue at least four different lines of scientific activity. The first in fact predates Caltech. From the very beginning, Max sought areas other than genetics that might allow the treatment of biological phenomena as problems of physics. While still in Berlin, he had met the great biochemist Otto Warburg who (among others) was working on a momentous effect of light on life, namely, photosynthesis. In photosynthesis, green plants convert carbon dioxide and water to organic matter (sugar) with the help of light; as a result, oxygen is released. This is the oxygen that higher forms of life breathe. Thus, in fact, light makes life possible as we know it.

Photosynthesis fascinated Max. He first learned about it in the private seminars at the Delbrück house in Berlin, and from Warburg, who in the 1920s had radically advanced the understanding of how light reacts with pigments. Warburg had done this by exploiting microorganisms, in particular, the ability of unicellular green algae to produce oxygen in light. With the aid of the manometric technique he invented, Warburg undertook the first detailed studies of the kinetic aspects of photosynthesis. He determined the first quantum numbers involved and established the general characteristics of the processes that depend on light (photochemical reactions) as well as those that occur independently (dark reactions). After Max left Berlin for Pasadena in 1937, he kept up to date on photosynthesis by taking a 1938 course given in Pacific Grove by Cornelius van Niel of Stanford University. In 1940, Max returned for a second round.

Van Niel's great achievement was to remove the photosynthetic process from a position of splendid isolation and to integrate it with other biochemical processes. He accomplished this by applying the principles of comparative biochemistry to reactions induced by light. In 1926, A. J. Kluyver and H. J. L. Donker had suggested a "Unity of Biochemistry" within a living system. Their major thesis proposed that the use of glucose by various organisms (dissimilation) consisted of a dehydrogenation process (oxidation) that was followed with the assimilation of this hydrogen (reduction) by intermediates. In 1931, Van Niel applied this proposal to green-plant photosynthesis, suggesting that the released oxygen is derived by dehydrogenation of a substrate (e.g., water), providing the reducing power for the fixed carbon dioxide.

In his teaching, Van Niel pointed out that "it is a pleasure to acknowledge a debt of gratitude to the microorganisms whose peculiarities have contributed to our better understanding of the photosynthetic process" (Kluyver and Van Niel, 1956). During the late 1930s, microorganisms were not used much in research. Van Niel's praise reinforced Max's preference for microorganisms. However, in spite of the inspiring teaching of Van Niel, Max never went beyond the role of a critical spectator in the photosynthetic field. He explained this in 1973 in a letter to Sir Hans Krebs, who had written a biography of Otto Warburg which Max had read "with burning eyes": "I found the Warburg style of controversy so disgusting and unprofitable that I never wanted to get into the field of photosynthesis, for that reason alone. Warburg had poisoned the atmosphere to an extent that it would have horrified anyone who came from Bohr."

A BIOLOGICAL LAW

While taking Van Niel's course in Pacific Grove in 1940, Max shared an apartment with Norman Horowitz, who was then a postdoctoral fellow. Their close friendship kept Max up to date in biochemical genetics, and later helped to bring him to Caltech from Nashville in 1947.

Beginning with his return to Pasadena, Max's experimental activities shifted from phage. He immediately began looking for a suitable system that would help in analyzing biological phenomena more complex than replication, just as phage had allowed him to study genetic mechanisms quantitatively. Specifically, Max was seeking the phage of vision. His first graduate student at Caltech, Roderick Clayton, began to analyze the response of microorganisms (purple bacteria) to light (D 43). Max eventually switched to a unicellular fungus named *Phycomyces*, picking it as his model system for analyzing the way an organism adapts to its environment.

The basic motivation behind this line of research was not only the complementarity argument. Max had something else very specific in mind: He wanted to discover a biological law, or more specifically to improve on a discovery more than a hundred years old, the so-called Weber-Fechner Law. This law stated that organisms that use light are capable of detecting intensity differences (Δ I) of about 1 percent over a large range. In other words, the change in the intensity (I) an organism can detect is proportional to the intensity it was adapted to. Expressed as a law in mathematical terms, this means that the ratio Δ I/I is a constant number.

Max described his interest in this biophysical law in detail in a lecture in 1949 (D 41):

> Textbooks of physiology assure us that excitability is a peculiarity of all living cells. From a physicist's point of view this feature may perhaps best be expressed in the following terms: a living cell is a system in flux: equilibrium, matter and energy taken in from the environment are metabolized and partly assimilated, partly degenerated, and waste products are given back to the environment. To a first approximation, that

in which growth is neglected, this represents a steady state. As long as the environment does not change, the cell does not change even though matter and energy flow through it. If the environment does change slowly, the state of the cell will also change slowly and continuously, but if the environment changes sufficiently rapidly the state of the cell will also change slowly and continuously, but if the environment changes sufficiently rapidly the state of the cell will also change abruptly. It will become excited. Thus, upon excitation a nerve cell will discharge an action current, a muscle fiber will contract, a sensory cell will discharge, and similarly all other kinds of cells under proper conditions of stimulation will produce a disproportionately large response to a relatively slight but rapid change in the environment. Do we here deal with a truly general feature of the organisation of matter in living cells? More than a hundred years ago Weber went one step further and formulated quantitatively a general law relating the threshold of stimulation with the parameters characterizing the environment. The law can be most easily formulated for situations in which one external parameter, say, the light intensity, changes abruptly from one level to another. Weber's law then states that the threshold change of this parameter which will produce excitation is proportional to the initial value of this parameter. . . .

The rating which this law of Weber has received in the biological literature has fluctuated from generation to generation, and has reached a very low level in recent times, so low in fact that it is barely mentioned in many of the current textbooks of physiology. . . . It remains to be seen whether closer quantitative studies of the law, particularly as displayed by the simplest organisms, will justify this low rating, or whether it will turn out that we have overlooked a powerful clue to the nature of the organization of the living cell.

Max hoped that a detailed study of this law that today is named after Weber and Fechner might become the nucleus of a growing science applying physical approaches to biological problems. Like the complementarity argument, the law reveals a simple solution for a complex situation. Max longed for biology to be governed by laws similar in kind to those of physics. The reason this hope was ill-founded resides in a basic difference between the science of matter and the science of life. For what appears simple in biology might be based on a variety of very complicated adaptive mechanisms that evolved over time. No parallel occurs for a simple physical principle. The intricacy underlying biological behavior results in "an ambiguity that pervades all of biology" (I). Thus the biological sciences cannot be constructed like physics; Max considered the term "biophysics" "ill-used," with little meaning (D 41).

A PHYSICIST LOOKS AT BIOLOGY

In 1949, when Max was invited to deliver an address at the one thousandth meeting of the Connecticut Academy of Arts and Science, he discussed this problem. He entitled his speech "A Physicist Looks at Biology." Since he chose a similar title twenty years later for his Nobel lecture in Stockholm, Max clearly considered himself "on permanent loan from physics" (Fleming, 1969). He wanted to understand biology as a branch of physics.

This becomes even more obvious when one considers the fact that in addition to the presentation mentioned above, Max offered quite a few more discussions of his view of the relation of physics and biology. By looking at Table 1, we can see that beginning in

Table 1
Max Delbrück's Lectures on Biophysics

Year	Place	Title	Occasion	Publication
1937	Berlin	"The Riddle of Life"	Departure from Europe	D 76
1943	Nashville	"Modern Biology in Relation to Atomic Physics"	Course at Vanderbilt University	Unpublished
1949	New Haven	"A Physicist Looks at Biology"	1,000th meeting of the Connecticut Academy of Arts and Sciences	D 41
1957	Cambridge	"Atomic Physics in 1910 and Modern Biology in 1957"	Supplementary lecture to Niels Bohr's Compton Lectures at MIT	Unpublished
1963	Copenhagen	"Biophysics"	50th anniversary of Bohr's 1913 publication of his atomic model	D 62
1969	Stockholm	"A Physicist's Renewed Look at Biology"	Nobel Prize lecture	D 76
1975	Pasadena	"The Complementarity Argument"	Last course on "Selected Topics in Biophysics" at Caltech	unpublished

1937 Max lectured or wrote about every six years on the connection between these two sciences.

The 1949 Connecticut lecture is of central importance. In a letter accompanying a copy of the manuscript that Max sent to Niels Bohr, he called it a "visionary lecture on complementarity." In the late 1940s, Max hoped to find a biological problem that required complementarity for its resolution. He organized discussion groups on complementarity at Caltech, inviting physicists to join him. One of the participants at this time was Aage Bohr, one of Niels Bohr's sons. Max had taught Aage Bohr how to do a standard plaque assay and the two became good friends, though they did not quite agree on the meaning of complementarity.

In the seminars, Aage Bohr received the impression that Max was overemphasizing practical aspects, that to Max complementarity applied to concrete experimental facts. He felt that Max seemed to expect complex molecules in living beings to have properties as alien to chemistry as the atomic spectrum is to classical physics. For example, the replication of nucleic acids might constitute such an alien property; investigating its mechanisms within the domain of physics might generate a complementary mode of description.

To Aage Bohr, the complementarity argument was more of a theoretical issue. He felt that one cannot say "what 'reproduction' is without characterizing it by concepts such as heredity, which are biologically defined," as he wrote his father in November 1949. Complementarity for Aage Bohr was operationally defined.

BIOLOGY AND PHYSICS

Max, however, continued to believe in a true observational complementarity; he hoped to encounter this in biology through a search for the limitations of the applicability of physics. Max was looking for an observation that could not be explained in straightforward molecular terms. He tried to describe his hope in concrete terms in the 1957 lecture given at the Massachusetts Institute of Technology in Cambridge.

One important change had occurred in biology since 1949, when Max delivered his address in Connecticut: The structure of DNA had been discovered. When Max first heard about the double helix, he felt that this genetic structure would advance biology in the same way Rutherford's 1911 atomic structure had advanced physics. But by 1957, he knew that the double helix had not brought biology to a higher level of understanding. For Max, this meant that no fundamental paradox had arisen:

> In 1910 . . . the program of physics seemed perfectly clear: more and better experiments, more refined techniques, more mathematical analysis, and the kingdom of atomic models will be attained. Today . . . we are at the height of enthusiasm for building molecules to interpret biological functions. To be sure, there are some disturbing features, but no point that could be pinned down, where we could say "Here is a manifestation of life, that could not rationally be reduced to an interpretation in molecular terms." Yet we have before us the example of physics, and might want to take a warning from it: complementarity might just be around the corner! To me this is an intriguing thought, and in fact the mere possibility of its being so can be one's principal motivation for an interest in biology. In practice it does not alter one's approach. The only avenue of progress in molecular biology today, as in atomic physics then, is to develop better techniques and to do more ingenious experiments, to drive molecular biology along its traditional and possibly naive path.

Max's lecture then went on to characterize the achievements of molecular biology in 1957 ("all this was undreamed of twenty years ago"); he pointed out, though, that a fundamental understanding had not been reached in biology. Max emphasized that biologists did not concern themselves to ask whether events such as recombination could be made visible as mechanical processes. Certainly the knowledge of the time did not allow such visualization.

Max's take-home lesson from the many examples he treated was that in 1957, there was "not a single elementary biological function to which we can give a coherent molecular interpretation." In analyzing his examples of muscle contraction and DNA replication, Max demonstrated that the desired reconciliation of these biological phenomena with molecular pictures works

> remarkably well as long as the description remains entirely on the formal chemical level, leaving out concentrations of the substances and geometrical aspects. As soon as structural aspects enter, that is, as soon as the reactions take place in particulates or on surfaces, tremendous difficulties arise. It is not to be taken as a foregone conclusion that structure on the molecular scale on the one hand, and integrated function on the other hand, are compatible observables.

He concluded with his hope that molecular biology might encounter complementarity in its pursuit of its goals, considering it "pleasant to think that further progress may demand from us a deeper analysis of the scientific method itself."

FRIENDLY SUGGESTIONS

Max's 1957 lecture was not greeted with enthusiasm by his friends. For instance, Wolfgang Pauli considered the examples "poor"; he missed a discussion of the truly deep consequences of complementarity which he thought could be found in fields such as cognition and psychology.

Max, however, deliberately avoided touching upon such topics, on the grounds that the issue had to be understood at the low molecular level first. Max not only looked at biological problems from a physical point of view, but he also *worked* at them. Such work does not begin at the higher levels of organization. Max remained firm in this, despite Pauli's criticism. Pauli believed Max's "emotional" interest in complementarity arose from the fact that he "shuddered at the thought, that life is 'nothing but a complicated wave function,'" as Pauli wrote to Weisskopf in 1958.

Pauli was not the only physicist who believed Max was thinking too small. In 1957 Niels Bohr himself found Max's way of expressing complementarity inadequate. Quite often, Bohr complained that the remarks he made on the relation of complementarity to life had been completely misunderstood. He kept on inventing new formulations, rephrasing his point of view until his death in 1962. A year later, Copenhagen commemorated the fiftieth anniversary of Niels Bohr's first paper on the constitution of atoms. At this meeting, during a special session on "Cosmos and Life," Max lectured on "Biophysics," introducing his talk as "the only one in which complementarity and the complementarity argument will be discussed seriously." Nobody else had been concerned with the subject, since "Bohr never permitted anyone else than himself to touch upon it."

In this 1963 lecture in Copenhagen, Max picked up on a topic that Bohr had discussed a year earlier at the inauguration of Cologne's new Institute for Genetics, which Max had helped to create. The point Max concentrated on in Copenhagen had not appeared in the printed version of Bohr's lecture (Bohr, 1963). As Max remembered, Bohr had spoken about the fact that in growth, "life is . . . confronted with a conflict between discreteness and continuity: we want to expand a volume continuously and we have to do it with discrete atoms of fixed size. . . . How can you expand uniformly if you cannot expand . . . the molecules?" (D 62). As examples Max discussed the synthesis of proteins and the replication of nucleic acids to analyze the tricks used by the cells in reducing "a three-dimensional continuous problem to a one-dimensional discrete problem."

In spite of the still unresolved details of DNA replication and protein synthesis, Max conceded in 1963 that there was "nothing peculiar" (that is, paradoxical) about these processes. Nevertheless, he kept his hope alive:

> The question remains, however, whether at the next higher level of organisation, for which we have no physical methods yet really, very important problems will come up not only here in molecular genetics, but in sensory physiology, in transport through membranes and other aspects. When we come to this next order of magnitude, there is the question whether or not something very peculiar from the quantum mechanical point of view, like superconductivity or superfluid helium, will come up. If strange cooperative phenomena can happen at room temperature in very special molecules, and if these can serve a special purpose, then certainly life will have discovered this 10⁹ years before we discover it. . . . Beyond that the question of whether our . . . present conceptual scheme: from information carrying molecules to proteins to small molecules making the building blocks for making the polymers,— whether that is a sufficient framework for accounting for all of biology . . . this is still a very wide open question.

After the lecture, his old friend Weisskopf rose to inquire if Max expected to find "an observational problem which leads to an observational complementarity." Max replied that "in the understanding of the next higher order," he hoped to "run up against an observational complementarity, so that we have to introduce different notions, I mean independent notions." When asked if he was thinking about "new laws of nature," Max responded: "I consider that that is possible and I am curious about it."

THE NEXT PHASE

Since there was nothing "peculiar" about genetics and its molecules, Max expected in 1963 that "the real issue of the relation of biology to physics is yet to come." He believed it was time to apply physical methods to study biological phenomena "at the[ir] next higher level of organisation" as physical processes. One of the phenomena Max had in mind was biological membranes.

In the 1960s, the principal device employed in molecular genetics was the famous double helix of DNA. From the point of view of physics, "the molecular biology revolving around this helix might be classified as one-dimensional molecular biology. . . . It stands to reason that nature, operating in three dimensions, long ago figured out that two-dimensional structures may also have their special virtues" (D 72).

Biological membranes are such two-dimensional structures. They are composed of proteins and lipids and exist as molecular bilayers of around 70Å thickness. These membranes serve an infinite variety of functions, which can all be controlled by environmental and chemical factors:

> This control is driven to its ultimate degree of discrimination in the display of surface specificity involved in development, and reaches its ultimate degree of sensitivity in the devices used to process incoming signals such as light, touch, or smell—devices used to adjust the behaviour of organisms to the external environment. Through modification of membranes the behaviour of the whole cell is then profoundly influenced. On the molecular level these transducer mechanisms are not understood and will constitute the principal challenge for the next phase of molecular biology. The depth of our ignorance in this area may be compared with the depth of our ignorance with respect to the molecular basis of genetics 30 or 40 years ago. We knew then that there were genes, and we knew that the genes were located in chromosomes, and we knew that they were arranged in linear order. We also knew that the chromosomes contained proteins and nucleic acids, but for several decades we thought that the proteins represented information storage (specificity, as it was called then) and that the nucleic acids represented a structural backbone. We now know that the reverse is true. Similarly, with respect to membranes, there exists an extraordinary degree of uncertainty as to the relative roles of protein and lipid: Which one determines structure and which one function? (D 72)

In order to attack this next phase of molecular biology, Max recommended the study of "Lipid Bilayers as Models of Biological Membranes" (D 77). In this approach, one forms lipid bilayers between two aqueous phases; with modifications of these bilayers, one can simulate some functions of real membranes. This research was pioneered by Paul Mueller and D. O. Rudin (1967, 1968), who succeeded in designing a clever set-up that allowed the construction of such a bilayer. The bilayer was formed by smearing a lipid mixture across a tiny hole in a partition separating two aqueous solutions. Under appropriate conditions, those lipid mixtures thin out to become "black membranes," truly bimolecular structures in which the polar heads of the phospholipids face the water phases, and the hydrocarbon chains form a somewhat disordered and liquid layer about 50Å thick.

One should appreciate the astonishing disproportions of such an artificial membrane: A bilayer 1mm in diameter and 50Å in thickness, if scaled up, would correspond to a piece of airmail paper 30 feet in diameter. These flat bilayers introduced by Mueller and Rudin are permeable to substances soluble both in water and lipid. The membranes are impermeable to most substances which carry a net electric charge such as anions (e.g., the chloride ion) and cations (e.g., the sodium ion). As a result, the electrical conductance of such artificial membranes is fantastically low. An unmodified lipid bilayer imitates the thickness, electric capacity, and the water permeability of real membranes. Max was convinced that work on these artificial bilayers would prove centrally important in attempts to understand the structures and functions of membranes in biological cells. Never lavish with approval, Max congratulated Mueller and Rudin on their "several very interesting discoveries."

Starting in 1969, Max began studying the electric conduction in these lipid bilayers. As particular additives acting on black lipid films, his laboratory investigated alamethicin (M. Eisenberg, J. Hall, G. Boheim) and cyclic polyethers (St. McLaughlin). Alamethicin is a cyclic peptide containing nineteen amino acids. It confers on lipid bilayers a conductance which varies over several orders of magnitude, depending on the voltage applied. By playing with the voltage or by using salt gradients across the black lipid film, one can simulate many striking electrical phenomena characteristic of nerve-axon membranes, including resting potentials, action potentials, and rhythmic discharges. But the molecular mechanisms involved remain obscure, and in spite of some great successes, the lipid bilayer technique never really contributed to an understanding of biological pores. Problems such as the acetylcholine receptor or the voltage-dependent sodium channel became attackable only after a new membrane technique was developed, the so-called patch-clamp technique (Neher and Sakmann, 1976).

Patch-clamp researchers succeeded in experimenting on functional membranes that showed specific responses to a specific and easily controlled stimulus. As early as 1965, Max had advocated investigating a membrane's response to specific stimuli (D 67). He had assumed this research would have to follow a different direction—isolating subcellular organelles as the functional membranes in mass and in pure culture. Development of the patch-clamp technique delighted him.

DIMENSIONS IN BIOLOGY

Max never became familiar with some of the complicated biochemical and physical techniques that were required to study "the next phase" in two dimensions. He preferred to think about membranes. Characteristically, he had learned about membranes by teaching a course on the subject. Beginning in 1951, Max had offered a regular course at Caltech entitled "Selected Topics in Biophysics." Like the phage course, this class continued for twenty-six years. It was listed as Bi 129, given in the winter term. Virtually every year he picked a new subject, spending an enormous amount of time learning the field. Max gave courses on chromosome mechanics, receptor physiology, and nonlinear differential equations. Mostly he talked about membranes. His first course on such a subject took place before anyone else at Caltech was paying any attention to membranes.

The cellular envelopes called membranes interested Max not only because they represented a step *up* in complexity, but even more because they represented a step *down* in dimensions, from three to two. The question of dimension becomes relevant when one examines how a signal moving in a cell can reach its target in time. Max was especially interested to learn whether a cell could use purely physical means for passing on a signal, that is, whether it could use diffusion in membranes. He explained his reasoning with the help of a mathematical theorem that was discovered by George Polya in 1921:

> Polya showed that the attempt to reach a given destination by random walk in unbounded space is certain to meet with success when it is

carried out in one or two dimensions but not so in three dimensions. In bounded space the counterpart to this theorem is: If one wants to reach a special site by diffusion in three-dimensional space, it is economical to embed the target site in a membrane to which the molecule in question will stick sufficiently tightly to stay adsorbed when it hits the membrane anywhere, and yet sufficiently loosely to enable it to perform two-dimensional diffusion on the membrane. It stands to reason, that nature started exploiting the implications of this theorem a few billion years before Polya discovered it. (D 72)

Apart from these general considerations, Max had become fascinated by the reduction of dimensionality when he had been working on a problem in insect sensory physiology. Max, who rarely attended scientific meetings, participated in the 1965 Cold Spring Harbor Symposium on Quantitative Biology that dealt with "Sensory Receptors." He became especially intrigued with chemoreception in the silkworm moth, *Bombyx mori*. A German group had presented the subject of insect olfaction (Boeck, Kaissling, and Schneider, 1965). They had shown that in the case of *Bombyx mori*, the particular odorant involved in chemoreception is "the sex attractant bombykol, exuded by the female and perceived by the male over great distances. The molecule involved is known precisely. . . . [It] is perceived by means of very numerous small sense organs, sensilla, each a little hair . . . innervated by the terminal processes of one or two sensory nerve cells, which run up through the length of the hair" (D 71). In order to exert their effect, the bombykol molecules have to reach these nerve cells.

What fascinated Max was the question of how they get there. This was indeed a problem since "over most of the length [the hairs] are shielded from direct contact with the outside air by a cuticular sheet. Direct contact occurs only at minute pores, 150 Å in diameter and spaced about 4500 Å apart from each other. Thus the pores constitute only about 1/1000 of the surface of the sensillum" (D 71). And the question was what happens to the 99.9 percent of the bombykol molecules that hit the hairs between the pores. In their Cold Spring Harbor presentation, the German group only discussed two possibilities: that the sex attractants are reflected, or that they stick wherever they land and stay put. They did not consider the possibility of diffusion.

Unsatisfied with the group's conclusion, Max discussed the problem with the experts at lunch (see Fig. 25). In one of the discussions, J. R. Platt pointed out a third possibility. In this version, one envisages that bombykol molecules, when hitting a hair's surface anywhere, will stick sufficiently loosely to be able to diffuse on the surface until they slide into one of the pores. This suggestion had already been put forward by Locke (1965) in view of the structure of bombykol and the probable surface nature of the hair. Such a regime would also gain high efficiency in catching odor molecules. Fascinated by the possibility that diffusion might be involved, Max immediately gave sensory physiology his attention.

At this time, Gerold Adam from Germany asked if he could do

25. *Max discusses chemoreception in the silkworm Bombyx mori with W. Pak (left) and J. Platt at Cold Spring Harbor in 1965. Reprinted by permission of Cold Spring Harbor Laboratory.*

some postdoctoral work at Caltech. Max agreed, telling him to examine the bombykol work done in Schneider's institute in See-wiesen and to acquaint himself with diffusion problems. Once Adam arrived at Caltech, he and Max collaborated on posing a theory that organisms handle some of their problems of timing and efficiency, in which small numbers of molecules and their diffusion are involved, by reducing the dimensionality in which this process takes place. They suggested a "new principle"—that the diffusing molecules reach their destination area not directly by free move-ment in three-dimensional space, but by subdividing the diffusion process into successive stages of lower spatial dimensionality. In the special application to the silkworm moth, the chemoreception was separated into two diffusion problems: first, the diffusion from the passing air to the surface, and second, the diffusion from a random point on the surface to a pore.

Max and Adam could show that finding a pore by surface diffu-sion works orders of magnitude better than do reflecting or purely adsorbing mechanisms. With this result, Max was willing to bet that surface diffusion is the natural mechanism, since "under these circumstances, one would be surprised if Nature had made no use of this possibility." This sentence by Dirac (1931) is quoted in Max and Adam's paper, entitled "Structural Chemistry and Molec-ular Biology," which they published in the Festschrift for Linus Pauling.

This was, in fact, one of the very few Festschrifts Max ever con-tributed to. For Gamow's Festschrift, he submitted a funny unpub-lished paper by Gamow with some amusing comments of his own. For Weisskopf, he submitted a discussion of Aristotle, whose works he was reading at the time. But for Pauling, he contributed an original scientific paper. Max had always felt sympathetic to Paul-ing's approach to science. Pauling's orientation was theoretical; he would first construct a theoretical framework, then consider exam-ples as applications. Max's contribution to the Pauling Festschrift followed the same approach. It presented first a general physical principle, moving on to solve a particular biological problem as an example of the theory.

Max always admired Pauling. When the great chemist received the Nobel Prize for Peace—his second—for his efforts to ban atomic weapons, Caltech did not respond with enthusiasm. Max had to step in and organize a party for Pauling.

A DEEPER LOOK

While Max was working on the signal transduction by diffusion, he had not forgotten about his "ulterior motive." In a 1968 letter to Warren Weaver, he wrote, "as to the future, I am still unwilling to let go my private fantasies. I am still waiting for the paradox though I may not live to see it come." The question has often come up as to why Max did not try to find the complementary situation he was hoping for by moving into neuroscience. There are several answers. First, from the very beginning Max felt science should climb the ladder of complexity step by step. Second, he felt that any analysis leading to clear paradoxes "should be done on the living cell's own terms" (D 41); one could do this most easily by investigating the information processing that occurs on the intracellular level. But any successful venture into neurobiology involved going beyond the intracellular level. One had to study how an outside signal could be converted to an internal one.

There was an even more important reservation concerning neurobiology. Max used the opportunity to present this argument in 1969, when he was awarded the Nobel Prize for Physiology and Medicine (Fig. 26). In his lecture in Stockholm, he grounded his argument on the assumption that studies in neuroscience eventually have to come to terms with the concept of truth. But truth, in Max's conviction, is a priori, and thus not amenable to scientific study. To justify his argument, Max referred to the Gödel Theorem (Tarski, 1969; Hofstadter, 1979), which states that in logical systems there are sentences which are true but which cannot be proven. He drew his conclusions from the fact that "the notion of truth, if it is to be meaningful at all, must be distinct and prior to the system of provable sentences, and . . . cannot be conceived as an emergent property of . . . a biological evolution. Our conviction of the truth of the sentence, 'The number of prime numbers is infinite', must be independent of nerve nets and of evolution, if truth is to be a meaningful word at all" (D 76). Max returned to this question six years later in his last course at Caltech, devoted to discussing the meaning of "Truth and Reality in the Natural Sciences."

The lecture in Stockholm was delivered in Swedish, in order to intimately involve the audience. When Max received the Nobel Prize, he was "happy"—as he cabled to Stockholm—but not over-

26. *Max receives the Nobel Prize in Stockholm (1969). Photo courtesy of Manny Delbrück.*

whelmed. He knew that what he had done was comparable to what others had done and he also knew that the choice of the winner is rather arbitrary (Fig. 27). He donated the money accompanying the prize to various charitable sources, and refused to make any more public appearances than he had before. He would not let the prize change his lifestyle. He did, though, seize the opportunity of a ceremony bringing together scientists and writers, to comment on the relation between arts and sciences. Max detected complementarity even there:

A Physicist's Renewed Look at Biology: Twenty Years Later
    Twenty years ago the Connecticut Academy of the Arts and Sciences had a jubilee meeting and on that occasion invited a poet, a composer and two scientists to "create" and to "perform." It was a very fine affair. Hindemith, conducting a composition for trumpet and percussion, and

27. Max dancing at the Nobel Ball. Photo courtesy of Manny Delbrück.

Wallace Stevens, reading a set of poems entitled "An Ordinary Evening in New Haven," were enjoyed by everybody, perhaps most by the scientists. In contrast, the scientists' performances were attended by scientists only. To my feeling this irreciprocity was fitting, although perhaps not intended by the organizers. It is quite rare that scientists are asked to meet with artists and are challenged to match the other's creativeness. Such an experience may well humble the scientist. The medium in which he works does not lend itself to the delight of the listener's ear. When he designs his experiments or executes them with devoted attention to the details he may say to himself, "This is my composition; the pipette is my clarinet." And the orchestra may include instruments of the most subtle design. To others, however, his music is as silent as the music of the spheres. He may say to himself, "My story is an everlasting possession, not a prize composition which is heard and forgotten," but he fools only himself. The books of the great scientists are gathering dust on the shelves of learned libraries. And rightly so. The scientist addresses an infinitesimal audience of fellow composers. His message is not devoid of universality but its universality is disembodied and anonymous. While the artist's communication is linked forever with its original form, that of the scientist is modified, amplified, fused with the ideas and results of others, and melts into the stream of knowledge and ideas which forms our culture. The scientist has in common with the artist only this: that he can find no better retreat from the world than his work and also no stronger link with his world than his work.

The Nobel ceremonies are of a nature similar to the one I referred to. Here, too, scientists are brought together with a writer. Again the scientists can look back on a life during which their work addressed a diminuitive audience, while the writer, in the present instance Samuel Beckett, has had the deepest impact on men in all walks of life. We find, however, a strange inversion when we come to talking about our work. While the scientists seem elated to the point of garrulousness at the chance of talking about themselves and their work, Samuel Beckett, for good and valid reasons, finds it necessary to maintain a total silence with respect to himself, his work, and his critics. Even though I was more thrilled by the award of the Nobel Prize to him than about the award to me and momentarily looked forward with intense anticipation to hearing his lecture, I now realize that he is acting in accordance with the rules laid down by the old witch at the end of a marionette play entitled "The Revenge of Truth":

"The truth, my children, is that we are all of us acting in a marionette comedy. What is important more than anything else in a marionette comedy is keeping the ideas of the author clear. This is the real happiness in life and now that I have at last come into a marionette play, I will never go out of it again. But you, my fellow actors, keep the ideas of the author clear. Aye, drive them to the utmost consequences."*

*The quotation is from Isak Dinesen's *Seven Gothic Tales* (Dinesen, 1934).

The Phage of Vision

"THE NEW LIFE"

In January of 1953, the discovery of the Watson-Crick double helix was only weeks away and molecular biology was about to reach its first climax. Max was kept informed about the progress in Cambridge by Watson's frequent letters, eventually receiving the first written communication about the DNA structure in March. Yet in January, Max's mind was occupied by a different subject. After several years of searching and testing, he finally had decided which organism he wanted to use to probe "higher levels of complexity" for complementarity.

On February 1, 1953, Max wrote to Seymour Benzer: "I am starting a new venture tomorrow, some experiments on the phototropism of the sporangiophores of *Phycomyces*. If they work, I'll retire from phage." This letter was typed in a jeep while "battling our way through the Sunday traffic." The Delbrücks were on their way back home from a four-day camping trip to Ensenada in Mexico, a "vacation before starting the new life."

A few months later, in the summer of 1953, Max gave a seminar at Cold Spring Harbor on "Phototropism in Fungi" (following Jim Watson's momentous talk presenting "The Structure of DNA"). The first experiments on the lower fungus *Phycomyces* had worked, and Max was convinced he had seized the phage of vision. The discovery of the double helix in the same year made his departure from genetics even easier. Geneticists were now going molecular and Max did not want to handle molecules, especially since there were plenty of good hands around to analyze chemical structures.

Apart from these technical and conceptional reasons, there were others. With all these advances in understanding the molecular basis of hereditary processes, it became more and more popular to work with phage and bacteria. One piece of advice Max frequently gave other scientists was "Don't do fashionable research." Max did not like to work in crowded fields with little space for new ideas. Crowds inhibit ease and spontaneity, which Max could not do without. He wanted to found a new group, and did not mind starting over again all by himself. On the contrary, he was "very happy with . . . [*Phycomyces*]. It is very satisfactory to work entirely by yourself for a while with not a person in the world sharing your interest," as he wrote to Luria early in 1954 (Fig. 28).

28. A lecture by Max at Cold Spring Harbor (1953). Photo by Seymour Benzer.

AN UNFAMILIAR FUNGUS

In his search for a new model system in the biological sciences which began in the late forties, Max was looking for a way to study the basic mechanism by which living things react to their environment. He considered this "one of the prime mysteries of biology" (D 41); it very likely still is today. As in genetics (and atomic physics), he felt, one should start with the simplest possible systems (phage and the hydrogen atom, respectively) in order to gain some insight into the means by which an organism perceives and responds to a change in its environment. The hope was that "by studying the responses of single cells to very simple stimuli we may throw some light on the behavior of more complex organisms" (D 41).

In Max's mind, the ideal system had to be a single cell that could be looked at individually by simple techniques. For that reason, after some preliminary trials with algae and bacteria, he eventually settled on the unicellular fungus *Phycomyces* as his phage of vision, his atom of physiology.

The life of *Phycomyces* begins when a spore is placed on some growth medium (potato extracts, for example). After a while, the spore will germinate and develop into a lawn called the mycelium (Fig. 29a). A few days later, little macroscopic stalks grow out of this fungal layer. When they reach about 1 centimeter in height, the stalks stop elongating and form little round heads at their tops (Fig. 29b). Nuclei from the mycelium are transported into each head, several of them being wrapped to form a spore. There are many thousands of spores in the sporangium, as the half-millimeter thick structure is called. The stalks that carry the sporangia are termed sporangiophores; their diameter is about a tenth of a millimeter.

After formation of the sporangium, *Phycomyces* resumes growing and the sporangiophores elongate about 3mm per hour. They can reach more than 10cm in length, forming a giant aerial single cell (Fig. 30). A particular region called the growing zone extends from 3mm below the sporangium. All the elongation is confined to this region, with the new cell wall material being incorporated here. While the sporangiophores are elongating, they are very sensitive to a number of stimuli. They respond to gravity, counteract stretching, bend into a wind, and grow toward the light. Max was fascinated by these responses. The growth continues long enough

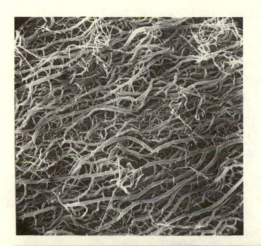

29a. *A scanning electron micrograph of Phycomyces mycelium grown on agar. Photo by Walter Schröder. Reprinted by permission.*

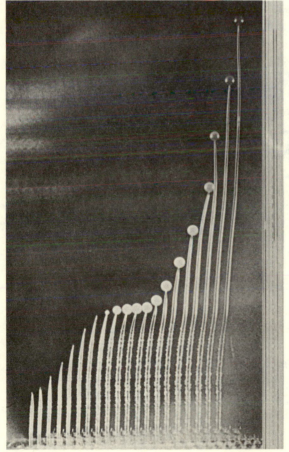

29b. *A developing sporangiophore, photographed at one-hour intervals. Reprinted by permission from Mannheimer Forum, vol. 72, 1972. Boehringer Mannheim; GMBH, Mannheim, West Germany.*

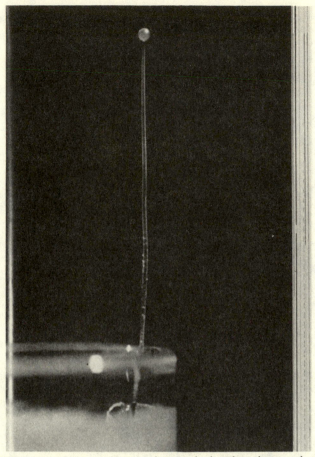

30. *The Phycomyces sporangiophore, or fruiting body. The sphere at the top, called the sporangium, elongates at the rate of 3mm per hour. This growth occurs in the region extending 3mm below the sporangium. Reprinted by permission from Mannheimer Forum, vol. 72, 1972. Boehringer Mannheim; GMBH, Mannheim, West Germany.*

to allow plenty of experiments on a single specimen in the course of a day; one could easily record all reactions by hand.

The behavior that interested Max the most was phototropism. If *Phycomyces* sporangiophores are exposed to light from one side, they will—after a few minutes latency—change their direction of growth and try to move the sporangium toward the light (Fig. 31). If *Phycomyces* is illuminated from both sides, it will transiently

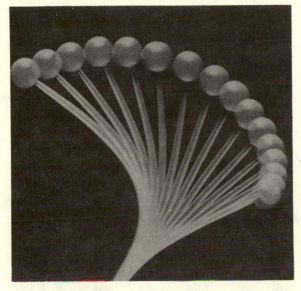

31. *Phototropism in Phycomyces (multiple exposures taken at 3-minute intervals). Initially oriented to the right, the sporangiophore was then stimulated by a light beam to the left. The sporangiophore bends about 3 degrees per minute. Courtesy of David S. Dennison, Dartmouth College.*

increase its growth rate but eventually adapt to the new situation and return to the original speed of elongation. This behavior is known as the light-growth response (Fig. 32). Though it does not serve any obvious purpose in the life of *Phycomyces*, it is the response best suited to scientific investigations. Functionally, the tropic response is the important one; it serves "the purpose . . . of seeking 'the open' so the spora can be widely dispersed. The light is used as a pilot to find this open space" (D 48).

RECRUITING FOR THE *"LILLE STILK"*

After committing himself to *Phycomyces* in the early 1950s, Max continued with it to the end of his life. It was a long-lasting love affair, fueled by the hope that the responses of organisms to a light stimulus could not be explained in physical terms alone. The fundamental property involved is the ability to be excited by some external stimulus. Although not a unique property of living matter, irritability, Max thought, was a fundamental characteristic of life. The fact that lifeless physical systems like neon tubes or Geiger counters also exhibit the property of irritability suggested to Max that the excitability of living tissues might be explained on a physical basis. He hoped that the essential relations between stimulus and response would prove the same in all excitable living systems.

Physiologists at this time were fascinated by all-or-nothing responses exhibited either by nerve fibers reacting to a current or by purple bacteria reacting to light. Max, however, felt "that threshold [all-or-nothing] systems represent a very poor choice, as far as simplicity goes." He suggested that "much more effort should be made to study graded systems. . . . In the search for a material more amenable to a thorough quantitative experimental study of the relations between stimulus and reaction, and involving a very pronounced range adjustment mechanism, we were attracted by the light growth reactions of . . . *Phycomyces*. This system has been studied off and on for almost 100 years, but we believe that its potentialities have never been fully exploited" (D 48).

Max wrote this after he had attracted the first scientist to join him in his new adventure. Werner Reichardt had met Max in Germany in 1954, when Max spent several months in Göttingen thoroughly combing through the older literature on biological stimulus-reaction systems. While in Germany, Max realized that he might have a better opportunity there than in America to recruit members of a *Phycomyces* group. In the mid-fifties, the new genetic gospel had not yet spread in Germany; few scientists there had heard of DNA.

Max got his chance when Bernhard Hassenstein gave a seminar in Göttingen on the system analysis he had done in collaboration with Reichardt in analyzing the ability of the beetle *Chlorophanus virilis* to perceive optical movement. Their goal had not just been to find any model to explain the data of the experiments; instead, they were out to find the simplest model possible (Hassenstein and Reichardt, 1953). In what Max praised to them as "a fundamental paper," they described how they planned to deduce the structures of the system under investigation by establishing relations between stimulus (input) and response (output). They were treating the beetle as a black box, using an approach advanced by the young science of cybernetics in the late 1940s (Wiener, 1947).

Max was impressed with Hassenstein's presentation. After the seminar they went for a walk. Max inquired about Reichardt, and when he learned that he was a physicist, Max determined to "get him to come to Pasadena and make a biologist out of him." Max visited Reichardt in Berlin, showed him what *Phycomyces* looked like, explained its responses to light, and told him what he wanted to study: "If you increase the light intensity, *Phycomyces* will speed up its growth transiently. After a few minutes it will be back to its

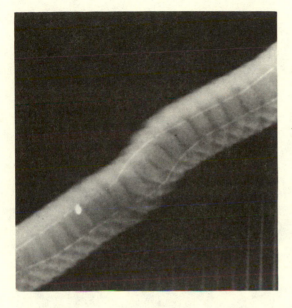

32. *Light-growth response of Phycomyces (multiple exposures taken at 2-minute intervals). A light pulse is given during the tenth exposure (note the highlight reflection spot). This causes a temporary increase in the growth rate after a delay of several minutes. Reprinted by permission from Mannheimer Forum, vol. 72, 1972. Boehringer Mannheim; GMBH, Mannheim, West Germany.*

normal growth rate. This behavior called adaptation is what I want to study and I want to attack this problem by system analysis. I want to improve on what Weber and Fechner achieved in the last century," as Reichardt remembered in 1982.

Reichardt didn't get a chance to answer right away, since Max had to leave to give a seminar. But when the two men met again later that year in Göttingen, Reichardt accepted an invitation to join Max at Caltech.

Before returning to Pasadena Max visited Copenhagen, where Niels Jerne, the now-famous immunologist who received the Nobel Prize in 1985, had invited him to lecture about *Phycomyces* at the Serum Institute (Fig. 33). Jerne remembers "waiting at the gate for Niels Bohr, who arrived a little late. With long strides he hastened in front of me, saying: 'I do not want to miss a word of what Delbrück says; I so enjoy listen[ing] to this man.' [Max] spoke in Danish about the 'lille stilk' that grew out of a 'lille urtepotte,' and refused to discuss the chemistry of the pigments involved" (Jerne, 1966, in PATOOMB).

In the discussion that followed Max's presentation, the questions concentrated on molecular details instead of on the behavior as a whole. This did not reflect Max's interest. After his return to

33. Party for Max in the Copenhagen apartment of Niels Jerne (September 1951). Max is in the foreground. Among the attendees are Ole Maaloe leaning back behind Max on the left, Niels Jerne standing behind Maaloe in tie, Gunther Stent in bow tie at the center, Andre Lwoff partially hidden in right rear. Photo courtesy of Gunther Stent.

Pasadena, Max once again explained his choice in a letter to Bohr on December 1, 1954:

> about Phycomyces, I am afraid during the seminar talk I failed to make clear my real ulterior motive. I talked about this system as something analogous to a gadget of physics, and explained at some length why it seemed more hopeful to me to analyze this gadget in great detail, rather than the many other biological gadgets which have been the subject of conventional research for many years. What I failed to stress was my suspicion, you might almost say hope, that when this analysis is carried sufficiently far, it will run into a paradoxical situation analogous to that into which classical physics ran in its attempt to analyze atomic phenomena. This, of course, has been my ulterior motive in biology from the beginning. What I have in mind is an application of the complementarity principle not in a form which is just vaguely analogous to the way it is used in physics, as having something to do with a shift in the dividing line between observer and object, but something much more closely related to the physics situation, springing directly from the individuality and indivisibility of the quantum processes.

ADAPTATION

In the late 1950s, about the time Max and Reichardt were collaborating, it was determined that *Phycomyces* phototropism, like human vision, is able to manage the enormous range of light intensities between twilight and bright sunlight (a range of ten billionfold in intensity). To make this possible, phototropism and vision, along with many other sensory processes, require sophisticated adaptation phenomena which allow the organism to adjust itself to the prevailing light level. Devoid of adaptation, an organism would not be able to handle the problems that life under the sun poses. The intensity of light at noon on a bright day is 10^{10} times higher than it can be on a moonless midnight. In a normal environment, the intensities vary from 20 to 1. Organisms that see have to be able to adjust to such a dynamic range, and the mechanism they employ is adaptation. As a consequence of this property, the sensitivity to light can be described by the Weber-Fechner Law. Stimuli have to be increased proportionally to the adapting intensity, to produce equal responses.

In order "to introduce a quantity which in some manner characterizes the sensitivity of the specimen at any given moment," Max and Reichardt introduced the concept of a "level of adaptation" (D 48). This rational measure is the inverse of the sensitivity. The level of adaptation is defined as being equal to the intensity of illumination with which the specimen would find itself in equilibrium.

Max and Reichardt could show that the light intensity (I) applied controlled this internal adaptation (A), such that a significant correlation existed between growth rate (output) and the so-called subjective intensity (i), defined as the ratio I/A. The aim of their joint "System Analysis for the Light Growth Reactions of *Phycomyces*" (D 48) was to create a "conceptual framework permitting us to describe the workings of the system quantitatively." They hoped the level of adaptation could become the appropriate parameter that described the gadget *Phycomyces*, as temperature or pressure characterize macroscopic physical systems.

Max liked to describe history in analogies. In *Phycomyces*, adaptation was the conceptional analogue of replication in phage; correspondingly, Reichardt would replace Luria in the realm of higher complexity blazing the trail. Unfortunately, after spending two years on *Phycomyces* Reichardt did not continue with it. He eventually resumed his work on the optomotor response of insects, because "its analysis seemed easy compared with trying to understand the influence of light on the dynamics of fungus growth" (Reichardt, 1966, in PATOOMB).

Reichardt was not the only one to have found that only the outward appearance of the unicellular *Phycomyces* is simple in comparison with a fly. Internally, a single cell must encompass complicated integrative actions. Here a fly can be simpler in its mechanisms, since it can distribute different tasks to different cells in various organs. This type of organization may be simpler to analyze than the organization incorporating many kinds of mechanisms in a single cell.

THE LIGHT IN PHYCOMYCES

One direction of research that Max pursued after Reichardt's return to Germany was to characterize the "visual pigment" of *Phycomyces,* that is, the "substance which undergoes a *photochemical reaction* which ultimately, through a more or less complicated chain of events, leads to the observable response" (D 55).

The modern term for "visual pigment" is photoreceptor. This designates the place where light meets life. Such an entity can be characterized by an absorption spectrum, if one can prepare a solution of the receptor. As long as the black box *Phycomyces* remained closed for biochemical approaches, one had to be content with more indirect means; thus Max initiated an in vivo characterization of the photoreceptor by a technique called action spectroscopy.

An action spectrum is the graphical representation of the sensitivity to light with respect to the wavelength. Historically, these spectra formed the foundation for understanding molecular details of photosynthesis; the master of this technique was again Otto Warburg. In a classical experiment in 1928 using yeast cells and bacteria, he had characterized the enzyme cytochrome oxidase, which is activated by light.

For the *Phycomyces* action spectrum, Max collaborated with Walter Shropshire. They had to be rather careful, since "the action spectrum *of the effect* may differ from the action spectrum *of the pigment* because of the intervention of screening pigments or of cell screening" (D 55). Max and Shropshire chose to alternate a standard light and a test light at five-minute intervals, giving a periodic output with a positive or negative amplitude, depending on the relative effectiveness of the light which is related to wavelength and intensity. By this procedure, they obtained an action spectrum showing four clear peaks. Unfortunately, it did not allow any conclusions as to the nature of the light-sensitive molecules involved; it did not even allow them to distinguish if a single receptor pigment was involved or a multitude. But Max concluded that the absorption spectrum of the photoreceptor(s) of *Phycomyces* should parallel the action spectra, and that this information could now be used to isolate the visual receptor.

A EUROPEAN INTERLUDE

At this stage, Max had to interrupt his work on *Phycomyces*. After long and difficult negotiations, he agreed to spend two years in Germany at the newly constructed "Institut für Genetik" in Cologne. Since the research in these laboratories concentrated on molecular problems, Max felt he could not take his favorite fungus along. Instead, he concentrated his experimental efforts on a chemical aspect of "light and life," investigating the "Photochemistry of Thymine Dimers" (D 58). His aim was to characterize in more detail the photochemical changes that occur in DNA upon exposure to ultraviolet (UV) light. The effect of photoreactivation was again on his mind, since the photodimerization of thymine had been observed to occur in DNA in its native configuration. The general suspicion was that ultraviolet light caused lesions in the genetic material. Upon his return to Caltech in 1963, Max maintained a strong interest in photobiology; he continued to investigate "the chemical and biological effects of UV irradiation on DNA and bacterial viruses," as he wrote in the 1965 Caltech Yearbook.

As the mid-sixties neared, Max felt he was drifting. He explained to George Beadle, "I am still fidgeting around to settle back into research that suits me, trying all kinds of things, from lunatic fringes to sober photochemistry. Have even pulled out *Phycomyces* again, but have yet to find the activity that really suits my temper. Perhaps I should train a duck to follow me around, that sounds like a very appealing way of life." After his return from Europe, Max found himself despondent for quite a few reasons. The first was his "longing for the charming lady Europe" *(die Sehnsucht nach der reizenden Dame Europa)*, as he wrote Jeanne Mammen a month after he had come back to Pasadena. Max noticed "worlds and a world history between this and that moon and the New World still appears to me stale and disgusting" *(fad und widerwärtig)*. Second, he had experienced health problems for the first time while in Cologne. Doctors had diagnosed his occasional chest pains as heart trouble. Despite his worry, he kept this condition to himself, for the first symptoms had appeared just after Manny gave birth to their fourth child, Ludina, in Cologne.

Thus, Max did not resume research at Caltech with full speed but spent considerably more time with the family. Besides, *Phycomyces* had not caught on yet. Ten years after Max started with bacterial viruses, the fields of phage and bacterial genetics were exploding. But ten years after Max began his work on *Phycomyces*, there was no similar momentum. Eventually, though, Max did pull *Phycomyces* out of his drawer to continue analyzing the behavior of this "humble mold." In order to harmonize the efforts on *Phycomyces*, he began a weekly group seminar in 1965; a year later, he termed these meetings "confession sessions." The first such confession session focused on mutagenesis in *Phycomyces* (April 28, 1966). In later versions, participants would commonly report on what they had done in the last week, hence the title (Fig. 34).

A NEW PHYCOMYCES GENERATION

Phycomyces was given new life in the late 1960s, mainly due to Enrique Cerdá-Olmedo, a geneticist from Spain. He convinced Max to apply genetics to study the fungus's behavior; he even introduced Max to the right chemical mutagen (nitrosoguanidine) and the proper conditions for moving in this direction.

Although *Phycomyces* had been one of the first organisms used in early attempts at artificial mutagenesis (Dickson, 1932), there were many technical and biological obstacles; only very scant results were obtained before 1966, for standard procedures produced no effect. The major handicap for obtaining mutants in *Phycomyces* was the number of nuclei in a spore. With an average number of about four nuclei, each equipped with a copy of a certain gene, normal morphology and behavior would be dominant. To Max, it seemed unlikely that any mutagen would work at all.

Then nitrosoguanidine was found to be effective not only in bac-

34. Max in his office at the California Institute of Technology. Courtesy of the University of Konstanz.

teria but also in higher organisms and even in *Phycomyces*. In 1966, Max was visited by Cerdá-Olmedo, who had just finished his doctoral work at Stanford University. Cerdá-Olmedo knew all the recipes for effectively making mutagenic cells with nitrosoguanidine. He found Max in San Diego, where the Delbrücks were spending a weekend at a favorite spot, the La Jolla Beach and Tennis Club. The procedure that Cerdá-Olmedo described seemed to Max eminently suitable for *Phycomyces*. Max immediately called his laboratory, where he reached an undergraduate (Ted Young) who was helping with a *Phycomyces* project. Max told Young what to do; when the Delbrücks returned to Pasadena, Max indeed found the first mutants. These exhibited colonial growth. Instead of covering a plate with their mycelium as normal *Phycomyces* does, the variants would clump in the form of thick colonies. The mutant hunt was now open.

After this Spanish injection, *Phycomyces* research gained vigor. A new generation of biologists joined Max and Cerdá-Olmedo. In the 1970s a fairly large *Phycomyces* group was at work in Pasadena, in Spain, and in Cold Spring Harbor where Max organized summer workshops (Fig. 35). Ten years after Cerdá-Olmedo convinced Max to study the mechanisms of cellular sensitivity and responses with the help of mutants, a "Behavioral Genetics of Phycomyces" had been established (Cerdá-Olmedo, 1977).

Concomitant with the new life of *Phycomyces* was the inception of a neurogenetic approach: the plan was "to generate single-gene mutations which affect behaviour, to analyse the mutants, and to try to infer general principles about how normal behaviour is created" (Quinn and Gould, 1979). Strangely enough, Max is credited with having opened this line of research, though he never took any active part in the question of how the physiology of nerve cells and the wiring patterns of neural circuits combine to produce behavior. Max did encourage behavioral genetics, though, once he was convinced of its feasibility, and he stimulated the interest in microorganisms.

BEHAVIORAL MUTANTS

The first treatment of *Phycomyces* with nitrosoguanidine produced mutants that were directly visible (colonial mutants). The second treatment also resulted in mutants of that sort, their color differing

35. *Phycomyces group at Cold Spring Harbor (Summer 1973). Photo by Manny Delbrück.*

from that of the standard strain, called wild type. These mutant strains were unable to complete the biosynthesis of β-carotene, which gives a characteristic yellowish color to the normal fungus. Since mutants are traditionally titled with three-letter words (Demerec, et al., 1966), the *Phycomyces* color mutants were designated by the abbreviation *car.*

Neither the colonial nor the *car* mutants were behavioral mutants; it remained a goal to find strains of *Phycomyces* that did not respond to light in the normal way. What was needed was an easy procedure to detect such mutants. Because one has to treat and investigate thousands of specimens to get to the right mutant, it is important to develop a procedure that avoids the necessity of handling each specimen individually. Martin Heisenberg, who spent two postdoctoral years in Pasadena, designed an elegant solution to the problem of isolating mutants for the light response. In this approach, agar plates with mutagenized spores were illuminated from below in a so-called glass-bottom box. In their search for the light, the sporangiophores of wild type grow down over the rim of the plate or wind along the surface of the agar; those of the mutants grow up, and their spores can easily be picked up. The mutants which appeared not to be attracted to the light, but which otherwise grew in a normal fashion, were named *mad*—a name suggested by Cerdá-Olmedo, who became Max's closest collaborator during the 1970s. Cerdá-Olmedo wanted to honor *Ma*x *D*elbrück with the name; Max, amused, accepted the term.

It was largely due to Cerdá-Olmedo that *Phycomyces* began to attract more scientists. He may very well be to this organism what Max was to phage. Cerdá-Olmedo is not only exceedingly bright, but dynamic. Max enjoyed collaborating with him, and even allowed Cerdá-Olmedo two-hour-long confession session solos. Such lengthy presentations by most others would have been curtailed (Fig. 36).

When Max started to contribute to *Phycomyces* again in about 1968, he decided that in order to get more scientists to share his new venture, a general review article was required. It was co-authored by the original *Phycomyces* dozen and published in *Bacteriological Reviews* in 1969 (D 74). Max picked this journal since it would provide a thousand free reprints, which he could mail all over the world. As a fungus, *Phycomyces* had nothing to do with the focus of the journal. Max was intending to depend solely on the

36. Max and Enrique Cerdá-Olmedo in the Delbrück garden (September 1980). Photo by Eliot Herman.

reprints to make the work known, as he had done with the paper on the nature of the gene in 1935.

THE AVOIDANCE STRUGGLE

The 1969 article introduces *Phycomyces* as "sensitive to at least four distinct stimuli: light, gravity, stretch, and some unknown stimulus by which it avoids solid objects." The avoidance response is described as the least understood of its sensory properties. These were the facts: "A sporangiophore placed close to a solid barrier grows away from it. The response begins about 3 min after placing the barrier 2 to 3 mm from the sporangiophore.... How the sporangiophore senses the barrier we do not know" (Fig. 37).

Even in 1986, the signal that *Phycomyces* perceives in order to start avoiding an object remains unknown; only negative evidence is available. For example, the response can occur in complete darkness, and the material of the barrier has no detectable influence. Both temperature and humidity have been eliminated as the avoidance-mediating signals, as have infrared radiation, or electrostatic forces producing a stretch response (D 90).

First described in 1881, the avoidance response was then forgot-

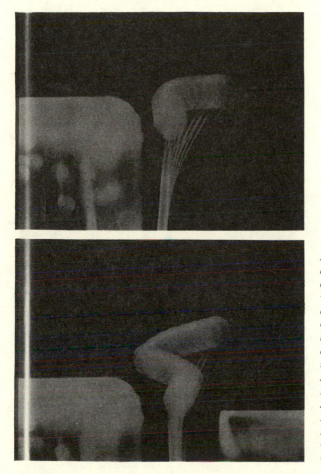

37. *Phycomyces avoidance response (multiple exposure, 2-minute intervals). The sporangiophore avoids the obstacle without contact. The sensing is believed to be mediated by a gas emitted by the sporangiophore, so there is a sort of chemical radar. Reprinted by permission from Mannheimer Forum, vol. 72, 1972. Boehringer Mannheim; GMBH, Mannheim, West Germany.*

ten for eighty-one years. It was rediscovered by Shropshire in 1962. Initially, it was only studied during summer workshops at Cold Spring Harbor (1965–68). When Max realized that the challenge was serious, he was "furiously working on avoidance all day" in Pasadena, as he noted late in 1970 in a diary he had been keeping ("one line a day") since 1964. His "furious avoidance struggles" did not bring an answer to the question: What is the signal detected by the sporangiophore that causes it to avoid an object? Two years later and ten years after the rediscovery of the response, Max felt "sick at heart at the unsolved state of the problem" (Diary, July 29, 1972). This is one of the few emotional comments to appear in the diary, which he kept to the end of his life.

Despite tremendous efforts, the problem remains difficult. It continues to intrigue, because the reaction is so simple and has great survival value for *Phycomyces*. In nature, the *Phycomyces* mycelium is likely to be found in dark, damp cracks. To disperse its spores, the sporangiophore has to find its way out. If there is light, the organism can use it as a cue. In darkness, it is the avoidance mechanism that frees the sporangiophore.

After considerable experimentation, Max and his coworkers settled on the "chemical self-guidance hypothesis" (D 75, D 90): "A volatile growth effector is emitted by the organism. The [artificial] barrier causes a concentration gradient across the sporangiophore. This gradient is sensed and causes the differential growth rate. Bilateral barriers result in symmetric changes in concentration, and hence cause a transient growth response." They further postulated that the barrier would affect the spatial distribution of the effector either by reflecting or adsorbing the effector molecules—thus altering the boundary conditions for diffusion—or by quieting local air currents. They hoped this self-guidance hypothesis would turn out remotely analogous to radar, with the signal pulsing being replaced in the chemical radar either by rapid decay or by strong adsorption of the effector.

In the search for the hypothetical effector, they exposed quite a number of volatile substances to *Phycomyces*. Most of these interfered negatively with the growth. Ethylene, however, did have the required positive effects, and is even emitted by the sporangiophores. Its rate of emission, though, is about one million times lower than would be expected if it were the effector substance. An effort was also made to detect the (postulated) effector by passing an air current through a "forest" of 1,000 sporangiophores and then onto an individual test specimen of *Phycomyces*. But no growth response occurred.

The physical nature of the avoidance signal has not yet been positively identified. Without evidence to the contrary, it is still postulated that the responses are all mediated by a volatile growth effector emitted and detected in the growing zone of the sporangiophore.

SIGNAL TRANSDUCTION PROCESSES

Part of Max's fascination with the avoidance response was conceptual in origin. He was looking for the paradox that would set limits to the reductionist approach which assumed that biological "stimulus-reaction chains" could be understood as physical stimulus-reaction systems (such as a doorbell). As early as 1957, Max stressed that "it is still true that none of the stimulus-reaction chains of living organisms have been fully interpreted" (D 82); the field was open for hopes and surprises.

In *Phycomyces,* the avoidance response seemed to present the awaited paradox right at the beginning of the transducing chain, where it would be most easily accessible. That such a chain can be readily seen in the case of *Phycomyces* was first noted by Cerdá-Olmedo and his collaborators in a paper Max termed the Three-Man-Work in *Phycomyces* (Bergman, et al., 1973). By bringing together the contributions of a physiologist (Kostia Bergman) and a geneticist (Arturo Eslava), Cerdá-Olmedo realized that one could summarize information about mutants with abnormal phototropism in a diagram relating the receptors to the responses. A classical genetic complementation test of these *mad* mutants (Ootaki, et al., 1974) provided the genetic background to the physiological diagram (Fig. 38).

By means of this diagram, one can directly examine a sensory transduction chain. The stimulus-response network of *Phycomyces,* however, should not be taken literally as drawn. Some of the genes may not specify actual sensory transducers, that is, molecules involved in the flow of information between stimuli and responses;

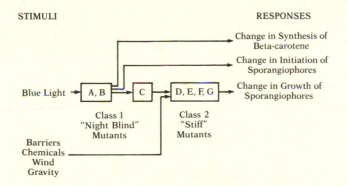

STIMULI   RESPONSES

Blue Light → [A, B] → [C] → [D, E, F, G] →

Change in Synthesis of
Beta-carotene

Change in Initiation of
Sporangiophores

Change in Growth of
Sporangiophores

Class 1
"Night Blind"
Mutants

Class 2
"Stiff"
Mutants

Barriers
Chemicals
Wind
Gravity

38. *A schematic representation of types of sensory responses affected by mutation of specific groups of phototropism mutants (the mad mutants). For example, the mad C mutant is defective only for growth modulation by light (phototropism and light-growth response), while being normal for all other responses. The A and B mutants are abnormal for all light responses; D, E, F, and G mutants are abnormal for all responses involving modulations of sporangiophore growth. After Bergman, et al., 1973.*

instead, their products may play an indirect role, such as physically interacting with these transducers or intervening in their synthesis.

From this scheme, the mechanisms underlying *Phycomyces* behavior do not appear forbiddingly complex. They are certainly not simple either, and it will take more than systems analysis to unravel the behavior of this primitive eukaryote. Physiological, genetical, and biochemical studies have to converge to advance the understanding of behavior.

MAX AT SEVENTY

Due to this success in joining genetics and physiology, Max's group attracted more collaborators; in the seventies, the *Phycomyces* group was truly international. There were graduate students from Taiwan and Germany, postdoctoral fellows from Germany, India, Spain, and Canada, and collaborating professors from Spain, France, the United States, and Japan (Fig. 39).

Max encouraged his group (and other foreign visitors) to "discover America"; he loved to teach them English, and to improve on their pronunciation. He would never speak German to the Germans. He even took away their English-German dictionaries, giving them Oxford dictionaries instead.

At Caltech, the *Phycomyces* group met about once a week for "confession sessions." Max even encouraged technical assistants to participate. Each speaker had to describe the background of his topic. If a presentation interested Max, it might last several hours, with Max frequently interrupting the speaker. Other topics he would gladly refer "to the subcommittee on biochemistry," quickly closing the session. Quite often at the end of a meeting, Max would invite the participants to a party at his home. On these occasions, he would entertain his guests by what he called "an instant puppet show." The Delbrücks had a small stage in their garden, with quite a few puppets. The basic cast included Max as the professor, and his daughter Ludina, then about fourteen years old. During the course of the party, Max would typically ask a few guests to join the puppet show. All participants would meet in a corner, agree on a plot, and then begin the impromptu performance.

In the mid-seventies, Max felt more than ever that *Phycomyces* had great potential. He himself spent a considerable amount of time in the laboratory, conducting his own line of experiments.

39. *Max and members of the* Phycomyces *group on camping trip in California desert (Spring 1974). Courtesy of E. Lipson.*

Though generally he left colleagues in the laboratory alone, if a particular project concerned him, Max would mercilessly push the student or postdoctoral fellow working on it. This was twenty years after he had first entered the world of *Phycomyces* and ten years after a fresh beginning. Though Max had been ready to give up on the fungus after the first decade, he was certainly not ready to do so after the second one. He was now willing to explore all possible angles.

Max explained his faith in *Phycomyces* when he turned seventy, albeit defensively. At that time he was up for retirement. Caltech waived its rigid retirement rules, offering Max the opportunity to continue active research at the institute. When asked what he intended to work on under these circumstances, he replied early in 1976:

> I feel that if I do make a serious experimental research effort (necessarily a very strenuous exercise) it should be in *Phycomyces*. I am still convinced that *Phycomyces* is the most intelligent primitive eukaryote and as such capable of giving access to the problems in biology that will be central in the biology of the next decades. If I drop it, it will die. If I push it, it may yet catch on as phage . . . caught on. Since I invested 25 years in this venture I might as well continue. I do not expect to make great discoveries, but if I continue to do the spade work my successors may do so.

Phage, intended as the atom of genetics, opened up the field of molecular biology. This success was due to the fact that there is a stage in the life cycle of a bacterial virus in which it consists solely of DNA. Furthermore, this phage DNA molecule can be isolated and subjected to a variety of chemical and physical studies. The story of phage is the triumph of the reductionist approach, though that is not what Max intended it to be. Max had been interested in replication, and was attracted by the fact that the phage molecule which replicated could be readily studied in the test tube.

In *Phycomyces,* Max was interested in adaptation and information transfer, but this is not performed by a single molecule. There is no simple reductionist approach to these problems, which is why *Phycomyces* remains a story yet to be completed. Phage work celebrated triumphs when hard thinking was complemented by the hard work of biochemical analysis. Unfortunately, all the biochemists' efforts on *Phycomyces* have thus far failed to significantly supplement the theories of geneticists and physiologists.

TRACKING PHYCOMYCES

The most obvious reason for the lack of a good *Phycomyces* biochemistry is the complexity of the behavior investigated. Even adaptation did not turn out to be a single chemical mechanism as originally envisaged. By more careful analysis using an automated tracking machine, *Phycomyces* was found to adapt to darkness in a different way than it adapts to light (Foster and Lipson, 1973; Lipson, 1975; Lipson and Block, 1983).

The tracking machine Ken Foster designed to precisely track and measure the growth of *Phycomyces* during programmed stimuli literally took the fungus out of Max's hands. Max had spent long hours in a darkroom recording the growth and bending of *Phycomyces*. He loved doing science with his own hands. In the early 1970s, however, the students protested, requesting more electronics. Eventually the tracking machine arrived. It required an extra room for the electronic-control units, the recorders, and the drivers. Max was never fond of computers. When shown the data-analysis computer one day, he smiled and asked, "Does this make people happier?"

A decade later, the tracking machine is in good hands, still being used to gain insight into the mechanisms of *Phycomyces*. It is now located in the Physics Department of Syracuse University, where Edward Lipson leads the largest American *Phycomyces* group, which is advancing biophysical research on the fungus. European work on *Phycomyces* mainly takes place in Spain and concentrates on genetic aspects.

Biochemistry still has a long way to go, although one breakthrough has been achieved from this direction. Even though for many years Max had refused to consider the chemistry of the visual pigment involved in the growth response, he did eventually become more and more familiar with questions about the nature of the molecule that receives the light. His final publication presented data that "strongly reinforced the conclusion that riboflavin is the chromophore of the photoreceptor" (D 110). It summarized six years of hard work, and Max was very proud of it.

LIGHT MEETS LIFE IN PHYCOMYCES

Since *Phycomyces* is most sensitive to blue light, its pigment is called a blue-light photoreceptor. Many plants are being investigated to determine the chemical identity of this light-sensitive molecule. Two candidates have been proposed as photoreceptors in *Phycomyces* due to their absorption spectra: β-carotene and riboflavin. Max first favored β-carotene because of this molecule's connection with the vertebrate photoreceptor molecule. In vertebrates, a molecule related to β-carotene serves as the light-absorbing molecule, rhodopsin.

In 1966, Gerold Adam disagreed with Max. After carefully comparing the *Phycomyces* action spectrum with the absorption spectrum of riboflavin, he suggested riboflavin as the responsible part of the receptor. Max, characteristically, did not believe a word of it. He even bet five dollars on β-carotene (which he never paid).

One difficulty in identifying one or the other as the blue-light receptor is that both molecules are involved in many other vital functions, thus proving ubiquitous at rather high concentrations. To decide this issue is then not so much like trying to find a needle in a haystack as trying to locate grass there.

The β-carotene was ruled out as the receptor after mutants were isolated containing only one thousandth of 1 percent of the amount of β-carotene found in the wild type. This concentration was too low to supply enough β-carotene to be used as the photoreceptor. But the phototropism of those mutants was normal, as David Presti, Max's last graduate student, showed in his thesis. With β-carotene now eliminated, riboflavin became the favorite candidate, thereby opening the gate for the road to success.

In 1975, Jose L. Reissig had suggested one should try to replace riboflavin in the photoreceptor with an analogue, a chemically similar substitute, and he tested several of them at that time. His argument was that if the analogue substitutes for riboflavin to form a functional receptor, one would expect that the peak of the normal wild-type action spectrum would shift toward the absorption maximum of the analogue. This brilliant idea ran into practical difficulties since the transport of the riboflavin analogues into the cell is handled by an enzyme which turned out to be subject to wide fluctuations in concentration. The gene coding for this enzyme was subject to frequent spontaneous mutations (D 105).

Max looked all over the world for help. The German chemist Adelbert Bacher suggested that he try a natural product, roseoflavin. This antibiotic, produced by a Streptomyces strain, has its absorption minimum where riboflavin has its maximum. A substitution would thus produce a considerable shift (toward the red) in the maximum of the wild-type action spectrum. Max was able to obtain some roseoflavin from Peter Hemmerich in Konstanz, who also synthesized dozens of other analogues for Max. None worked suitably. Hemmerich aided Max in obtaining more roseoflavin, which was subsequently provided by K. Matsui in Japan.

In 1978, Max invited Manfred Otto to try replacing the riboflavin. Otto had been working in Bacher's laboratory in Stuttgart. He first ascertained that roseoflavin is taken up by *Phycomyces*. By analyzing the cell's flavin metabolism, Otto determined the maximum concentration of roseoflavin that did not interfere with the phototropic response. In order to maintain a constant level of flavin in the cell, Otto isolated *Phycomyces* mutants that did not synthesize riboflavin. All their flavin had to come from the growth medium.

The crucial experiment involved checking the phototropic balance, an idea suggested by Reissig back in 1975. A sporangiophore is placed between two light sources. One emits blue light (the maximum absorbed by riboflavin and the minimum absorbed by roseoflavin), the other green light. If both sources emit light of the same intensity, wild-type *Phycomyces* would grow toward the blue light.

By using a higher green-light intensity, one can find a position at which normal *Phycomyces* is in balance, so that it grows straight. Then the strain that has been grown on a medium containing roseoflavin is placed at the same spot (the balance point). If roseoflavin has replaced riboflavin in the photoreceptor, the strain should bend toward the green light. This is just what occurred (D 110), leading to the conclusion that the riboflavin analogue, roseoflavin, did act as the visual pigment in the mutant and thus that riboflavin is part of the photoreceptor.

This success did not clarify everything. It left the question open as to how the light is detected and how the light signal is converted and transferred to the enzymes that eventually take care of the growth reactions. In addition, the interpretation of the blue-light photoreceptor is now becoming more complicated. Work in Lipson's laboratory in Syracuse suggests that there are two such receptors that enable *Phycomyces* to adapt to a huge range of light intensities and thus remain sensitive in the same range as the vertebrate eye. Since the human eye makes use of two structures (rod and cones) for handling high and low intensities, it would not be too surprising to discover that *Phycomyces* also employs two molecules to do the job.

Lipson and his group are working to identify photoreceptor mutants and to interpret all the genes involved in the light transduction chain. He is the only physicist recruited by Max for *Phycomyces* work who has stayed with it. Recent results suggest that Lipson might indeed get his hands on the mechanism of adaptation, just as Max envisaged when he began with *Phycomyces* more than thirty years ago.

The German Interlude

RETURNING TO GERMANY

While the world outside was fighting the war, Max was preparing the foundations for the new genetics in a quiet laboratory in Tennessee. As soon as the war was over, though, he began to look for ways to help his German friends and relatives. He also immediately devised plans to get to see them. The first visit was arranged for the summer of 1947, which he intended to spend mostly in Copenhagen. Since the State Department would not validate his passport for Germany if he went for personal reasons, Max asked H. J. Muller, then president of the Genetics Society, if he could send Max as "an emissary at little or no cost into Germany." In a letter to Muller on January 25, 1947, Max outlined his plans for how he could get to Berlin in spite of the complicated fragmentation of the country into four zones:

1) Obtain a request in this country from a scientific society to do a job for them in the British zone . . . of Berlin,
2) While passing through England, obtain from the British an authorization to enter their zone . . . and proceed to Copenhagen,
3) Obtain in Copenhagen from the American Consul validation of my passport for Germany, and proceed from there by Danish airline into the British zone . . . and having accomplished my mission there, proceed by British airline to Berlin,
4) Obtain in Berlin permission to go from the British sector . . . to the U.S. sector. . . . This last step one can do during the daytime without specific permit . . . but not at night.

Muller agreed to have Max represent the Genetics Society of America in Germany, asking him on behalf of the Committee on

Aid to Geneticists Abroad to collect some firsthand information as to which geneticists were still in Germany and what evidence there might be concerning whether or not "they may be regarded as innocent of having actively aided Nazism and taken part in the Nazi prostitution of genetics." Muller also inquired about addresses of families, since the committee could furnish a number of CARE packages.

Max promised to make an informal inquiry into the private and public lives of German geneticists over the past ten years (which he actually never did, although he provided some addresses). In July 1947, he left New York on a Danish freighter for Copenhagen. The three weeks in Germany depressed him: "all the railroad stations were still beleaguered by thousands of people who were just camping out, and very little transportation, and food and everything. The currency hadn't been reorganized yet" (I).

The war had produced an enormous amount of physical destruction, making the situation in Germany chaotic. The Delbrück house had been completely destroyed. Max's mother had died the year before: it was comforting that she had not seen the loss of the house she had lovingly planned and enjoyed with her family.

While in Berlin, Max felt guilty for not having stayed in Germany; he respected his many friends who had remained and had tried to prevent or rescue it from disaster. The scientists he admired who had stayed on in Germany included Karl Friedrich Bonhoeffer, Hans Kopfermann, Max von Laue, Werner Heisenberg, and Otto Hahn.

Everywhere, Max observed a sense of relief that the war was over and that rebuilding could start. He immediately conceived of an effective way to help scientifically, offering to send Max von Laue all post-1940 copies of *Physical Review*. Max assumed, correctly, that the Kaiser Wilhelm Institutes would still have issues up to 1939. Von Laue was deeply moved and offered to help Max during his visit. Max in return asked Von Laue to find people to work on phage.

For more than thirty years thereafter, Max continued to help German scientists. In 1967, he mailed eighteen boxes to Erhardt Geissler in Berlin-Buch (East) filled with the 1955–67 editions of the *Proceedings of the National Academy of Science;* he promised more, hoping "they don't give you indigestion." Until his death, Max would mail his own personal copies of the *Proceedings* to Berlin.

This mission was then assumed by Pat Burke, who had come to Max as a *Phycomyces* student and became a close friend of the family.

During his first visit to Germany in 1947, Max visited Otto Hahn in Göttingen. Hahn was then the president of the Kaiser Wilhelm Gesellschaft (which in due course became the Max Planck Gesellschaft of today), and was devoting his efforts to preventing the dissolution of the organization, as had been decreed by the four-power control commission. It was a depressing job, in a confusing and uncertain situation.

Most of the institute buildings had suffered physical destruction during the war. One structure to escape unharmed in Berlin was the Harnack House, named after Max's uncle, the founder of the Kaiser Wilhelm Institute. It had been used by the faculty club as a lecture hall before the war. During the war, the American officers' club took it over, thereby bestowing the ultimate insult on many in Germany. Otto Warburg was quite pleased to use the occasion of Max's visit to Berlin to return Harnack House to scientific purposes. He arranged for his American visitor to give a talk there. Warburg "arranged it, I think, [for] 2:30 in the afternoon, and I said, 'Well, 2:30 in the afternoon is a pretty bad time. I mean German professors have a good midday meal, and then they need a siesta. They wouldn't be able to stay awake.' 'Oh,' he said, 'they don't get that much to eat these days that they can sleep afterwards' " (I).

BRINGING PHAGE TO GERMANY

Warburg insisted that for the occasion of Max's seminar, Germans must again be admitted to Harnack House. Thus two young students from Berlin saw the electron micrographs that Max showed of his favorite phage, T2. The students were Carsten Bresch and Wolfgang Eckart, both of whom wanted to specialize in genetics. Postwar Berlin did not offer many opportunities, and they had been forced to work on trout and carp. When Max showed the phage, Bresch decided they should forget about the fish and learn more about phage.

The two students convinced their advisor, Robert Rompe, to write a letter to Max after he was back in the States. Max immediately sent phage to Berlin, which he announced in a separate letter; the samples were, however, held up by an American control officer.

After a while, when nothing arrived, Bresch investigated, only to find Max's package in an office. Though the agar looked rather crumpled, the bacteria were still alive, as were the phage particles. Thus T phage had arrived in Berlin. When Bresch and Eckart began working on it, they had very primitive equipment and only a few hours of electricity a day; since their water bath needed an hour simply to reach the required temperature, they couldn't do any complicated experiments. But their enthusiasm overcame such hindrances.

ANOTHER SUMMER IN GÖTTINGEN

Max visited Germany regularly in the following years, but it was not until 1954 that he stayed for any length of time. In that year, he went to Göttingen for a few months as a guest of the Max Planck Institute for Physical Chemistry, where his old friend Karl Friedrich Bonhoeffer was the director. Max gave quite a few seminars about the new genetics and the double helix. Germany had not contributed to the germinal developments that gave rise to molecular biology during the war years and in the immediate postwar period. The double helix was news in Germany in 1954. Werner Reichardt had never heard of it, nor had the great Otto Warburg.

The seminar Max gave in Göttingen is well remembered. Max was solemnly greeted by the dean of the Natural Sciences Faculty, who wore his official robe for the occasion in spite of the extreme summer heat. Max, dressed in sneakers and a T-shirt, opened his presentation by announcing that owing to the heat, he would first show all the slides, then open the window and continue with his talk. He concluded his seminar by praising the lifestyle at American universities, where he did not have to tuck his shirt into his trousers. Max always admired the other side of the fence. In Germany, he would praise American customs, while in Pasadena he would applaud German habits and culture, opposing any negative remarks.

The Delbrück family returned to America by boat that year. They embarked with Niels Jerne on a Dutch freighter that brought them from Antwerp to New Orleans. Besides playing chess and Ping-Pong, Max and Jerne joined in playing a practical joke. They had noticed that at noon each day, the crew hauled up a bucket of seawater along the side of the ship to measure the ocean temperature. Max and Jerne managed to place in the bucket a sealed bottle containing a piece of paper bearing Arabic script. Later in the evening, they enjoyed listening to the captain develop fanciful theories to account for the origin of the bottle.

THE COLOGNE CONNECTION

One of the scientists to attend Max's German seminars in 1954 was Joseph Straub, of the Department of Botany at the University of Cologne. He immediately became convinced that the new genetics represented the scientific future and that it should be carried on at a German university as soon as possible. That summer, he sent a letter to Max asking him to suggest candidates for a department of microbiology to be set up at Cologne. In response, Max suggested Wolfram Weidel and Gunther Stent.

Straub persuaded the faculty that the new professor should pursue genetics research. He also convinced them to offer this position to Max. A letter was mailed to Pasadena, informing Max that the faculty list of proposed candidates ran as follows: (1) Delbrück; (2) Weidel; (3) Stent. Max answered with his usual postcard, declining the position. Straub responded with a request that Max wait for the official offer before reacting. When it came, Max promptly turned it down.

In the meantime, though, the second person on the list had been captured by the Max Planck Institute. Max, feeling that Straub needed some help so as not to lose the momentum he had built up, offered to come for three months to give a course on phage genetics. This eventually took place in 1956.

Straub remembers well quite a few special circumstances of the Delbrücks' 1956 visit to Cologne. He picked up Max and Manny and their two children, Jonathan and Nicola, at the airport in his VW Beetle and brought them to the house he had rented for them. Max and Manny then went shopping while Straub was left in charge of the children, who were eight and six years old. In the living room of the house, the beautiful armchairs seemed to invite Jonathan and Nicola to play hide-and-seek. Finally the armchairs were used as trampolines; Straub feared the worst, especially when the owner of the house showed up to observe the proceedings. When Max and Manny returned, they were told what had transpired. Max dragged his children in front of the landlady and explained to them in

German (which they didn't understand) that Germany is famous for the discipline of its people and that there is a holy law about armchairs, which are meant only for sitting in, and then only in clean clothes. The landlady was convinced she had the best possible tenant.

After this incident, Straub took the opportunity to say good night to the Delbrücks, only to hear Max inquire when the phage course would begin. Straub had hoped Max would wait a few days before asking about the preparations for the course, since the building Max was supposed to teach in was just being completed. The water baths that Max required were still under construction. When Straub described the situation, Max announced nevertheless that he wanted to start in two days. Straub put some night shifts to work to get the equipment ready; the course did start on schedule.

In the summer of 1956 (and also in later years), Max introduced in Cologne the same integration of personal and scientific family life he had been noted for in Pasadena. He encouraged parties, and along with Manny, contributed enormously to social life. The favorite activity was a treasure hunt. Max would distribute letters containing puzzling clues as to how to find a treasure hidden in the university area. Usually the "treasure" consisted of a bottle of local beer, which Max was fond of.

THE PLANNING OF AN INSTITUTE

One of the participants in Max's 1956 phage course was Straub himself, then dean of the Natural Science Faculty. One day, he had to get to his office by 11:00 A.M. to conduct his official duties. He asked Max for permission to leave. Max declined, insisting, "You stay here."

And Straub did. For his long-term plans, he needed all the time he could get with Max. Over and over again, Straub told Max how impressed he was by the developments in genetics and how strongly he felt that Germany had to get back into the contemporary mainstream. He used every opportunity to promote his idea of a new institute for genetics at the University of Cologne. Several times, he asked Max to name conditions under which he might be interested in organizing and directing such an institute.

Max felt Straub's idea was on the right track for getting German (and European) university biology going again. Characteristically, the development of modern biology was done by outsiders trained in physics, chemistry, or medicine, using organisms such as *Neurospora*, bacteria like *E. coli*, or phage. Research in molecular biology was simply not possible within the classic German separation of biology into botany and zoology. Furthermore, Germany in those days showed little activity in genetics and its biochemistry. As these fields developed abroad, all the available German faculty positions were already occupied by botanists and zoologists, and physiology and microbiology were taken up by medical schools, which focused on therapeutic interests rather than on general biological problems. In 1956, three years after the DNA structure had been elucidated, there was hardly a professor of genetics, biochemistry, or microbiology anywhere in Europe. Thus Max was eager to further the state of biology at German universities; yet he considered Straub's plans impossible to realize.

Straub, however, turned out to be a superb negotiator, using Max as a lucky charm. He lured Max by telling him the administration was going to act, and he lured the administration by saying that such action was a condition for Max's coming. Eventually he was told by the president of the university that it might indeed be possible to finance an institute for genetics, "if Max Delbrück was willing to run it."

This opinion was probably strongly influenced by the university's chancellor, Friedrich Schneider, who had been impressed by Weidel's book on viruses (Weidel, 1956). Schneider arranged for a meeting between Max and the university's president, Professor Kaufmann. For this occasion, Max bought a tie. After the opening small talk, when the president asked Max to provide a quick tour of the phage world, Max sat back, folded his hands behind his head, looked at the ceiling, put his feet on the president's desk, and thought for some time about the question. Neither said a word. Eventually, Max described what had been achieved in genetics, outlining future projects.

Straub eagerly seized on Max's interest "to carry modern biology right into a university setup," (I) immediately arranging to discuss the situation with representatives of the government and of the funding source, the Deutsche Forschungsgemeinschaft. The proposal to create a new institute in Cologne now appealed very strongly to Max, especially since he felt "very great confidence in the abilities of Straub to carry off anything he proposes to do," as he wrote to Beadle two days after his return to Caltech in the fall of 1956.

In this letter, Max summarized all the discussions Straub had arranged; he also outlined the conditions he had named for coming to Cologne:

> I said that I would be interested if something could be created which could serve as a model for other universities in Germany and in other countries for breaking down the organizational deadlock in which biology finds itself all over the world, with Caltech almost an unique exception. The deadlock I am referring to is . . . the fact that departments of Botany and Zoology were created long ago, and that their creation . . . has stunted the growth of genetics. . . .
>
> The following plan was then outlined: there should be created at Cologne an Institute somewhat along the lines Morgan drew up for Kerkhoff when he came here: a lab building to be constructed in two steps, work on the first half to begin immediately and to be moved into in two years. The first half . . . to be staffed with a director and five "professors" (in the American sense), and further suitable staff. . . .
>
> The consensus of the people who took part in these discussions was that this was a plan that would be financially feasible, and that the next couple of months would be particularly favorable for pushing through such a plan, both for a building grant of DM 2.000.000 from the Deutsche Forschungsgemeinschaft, and for appropriations from the parliament of the Land Nordrhein-Westfalen for getting the Institute going and for setting up a continuing budget. All this, however, provided I would make a commitment to be the initial director of this Institute, if and when it comes into being.

Max wrote this letter two days after his fiftieth birthday. In it, he also inquired of Beadle as to "the possibility of obtaining a two-years' leave of absence" from Caltech. Max explained that he did not want to stay in Germany for more than two years because "under the present U.S. laws I would lose U.S. citizenship upon return to my original country for a period extending over more than two years." The letter described the appeal of the Cologne challenge mainly as "the 'attraction of a vacuum,' i.e. to be unencumbered by a preexisting organization."

In spite of great efforts, it took Straub another year before he had worked out all the details. In November of 1957, he applied for money from the Deutsche Forschungsgemeinschaft to construct an "Institut für Genetik" in Cologne. His eleven-page, single-spaced letter is a masterpiece. Largely since he could report that Max was

ready to become the first director (for two years), the Forschungsgemeinschaft approved the plan. After five years, the institute finally was completed. Max and his family were in attendance from
1961 to 1963.

The years prior to 1961 were filled with complicated negotiations
about details such as the number of floors in the building and the
length of fire escapes. Some of the problems raised serious philosophical issues. For instance, the architects added a special feature
to a drawing that Max had made. He complained about it: "I don't
like the large 'Director's Office' sign on the Library floor. I want to
emphasize democracy. What I have in mind is to have a group of
five people of practically equal standing with one (possibly rota-
ting) chairman."

Max wrote this in a letter to Carsten Bresch, who had already
joined the faculty in Cologne in 1956. He was later joined by Walter
Harm. Max also brought Peter Starlinger to the institute. He began
as an assistant, developing into the most important geneticist at the
institute Max built.

THE YEARS IN COLOGNE

The Delbrücks left Pasadena at the end of July 1961 in a big bus
some of their friends had chartered for the ride to the airport. With
more than a dozen pieces of baggage and twice that number of
friends, Max and his family traveled along the freeway to the airport accompanied by singing and bagpiping. The bus took them
right up to the plane.

That joyful mood did not continue in Cologne. Here they found
a seven-story institute which had been scheduled to be fully operating in April, still with elevators that did not begin to function until
October and a heating system still only promised. Fortunately,
Max's sense of humor accompanied him to Germany. His first report to Caltech's *Biology News Notes* concerned the display of ele-
gance among German ladies. Max evolved the following rule: The
shorter the skirt, the taller the coiffure.

In Cologne, Max oversaw the construction of an institute designed to promote collaboration. Twenty years before that, such an
interaction would have been the exception. Each professor would
have jealously protected his own special field, preventing others
from interfering. Interdepartmental cooperation was an alien con-

cept for German faculty; eventually faculty members such as Bresch and Harm did become convinced of Max's conception.

What people remember most of Max's stay in Cologne was that he invited prominent American scientists to speak at the institute; stimulating discussions always followed, during which Max encouraged young students to speak up. Max himself used the years in Germany to introduce the new developments in genetics and to outline his approach to *Phycomyces* (D 60–D 67). In addition, he maintained his contact with Niels Bohr. Since Cologne was so close to Copenhagen, Max invited Bohr to deliver the opening lecture at the new institute. Max challenged him to reconsider the arguments he had presented thirty years ago (Bohr, 1933), in light of the rapid developments of molecular genetics in the interim and in view of the fact that "the problem of being able to interpret phenomena of life by molecular mechanisms is not just a topic of conversation but a daily scientific question."

On May 6, 1962, Max suggested that Bohr call the opening lecture "Light and Life—Revisited." Bohr accepted with enthusiasm (Bohr, 1963). He came to Cologne intending to give the lecture in English, but learned upon being introduced to the deans and the president of the university the night before his speech that they could not speak English. So Bohr decided to switch to German. No one in the audience seems to have understood a word anyway.

Those who met Max while he was in Cologne tend to recall two impressions. Many were overwhelmed by his wide-ranging interests and his ability to state complicated lines of reasoning in a classical, simple way. Others despised his manners. Max upset them by his idiosyncratic behavior, such as walking out of seminars.

At the same time, he is remembered for his love of games. For example, Max liked to play Scrabble. Of course in the Delbrück house, the game was played in an unconventional way, sometimes using phonemics. Newly created words were permitted if an established pattern was used. One night Max produced the monster word "Neunschirmyogi." He explained this would be a yogi who performs with the help of nine umbrellas (*neun Schirme* in German). When Hansjacob Seiler, a professor of linguistics in Cologne, challenged this construction, Max insisted, promising a detailed proof in due course. A few days later, Seiler was invited to join "the solemn inauguration of a painting, that is the exact picture of a

Neunschirmyogi." Seiler saw an oil painting by Jeanne Mammen that exhibited nine umbrella-like structures. It is now officially catalogued under this title.

Another of Max's favorite games was Hinkel Finney Duster, which he had learned from a former Russian girl friend. It's played with a deck of cards that includes the king, the queen, the knave, the 10, the 7, and the 3. Together they form a family. King and queen are father and mother, and the knave is brother Jack; each family has ten daughters, seven sons, and three dog treys. The goal of Hinkel Finney Duster is to gather a complete family. The four families are called the family of the heart's delight (hearts), the jeweler's family (diamonds), the gravedigger's family (spades), and the policeman's family (clubs). One can get a card (or a member of a family) by asking another player for it. "Do you have the ten daughters of the family of the heart's delight?" If the answer is "No," the one who was asked continues. If the answer is "Yes," the desired card is placed on the table. The important rule is that before picking up the card, one has to say, "Thank you."

The Delbrücks insist that each time they played this game someone forgot to say, "Thank you," especially someone close to completing a whole family. The forgetter would have to throw his cards in the air, putting them up for grabs. A wild scramble ensued.

COLOGNE AFTER MAX

While he was there, Max dominated the institute in Cologne and his personality was strong enough to prevent any fighting. But his departure in the fall of 1963 left a vacuum. Problems continuously were fielded for him to solve in Pasadena. One letter from 1964 complained: "You should tell some of the younger colleagues that running around without a tie does not mean one is a Delbrück." The institute ran into a crisis, mostly because the faculty could not make up its mind whether to appoint Bresch or Harm as full professor; in fact, Max had not given any clear indication as to how he felt. Eventually both moved on, as did Heinz Zachau, who had joined the institute a few years earlier. This left only Starlinger, who is still there today, joined by a new generation of geneticists.

In a letter on May 19, 1964, to the dean of the faculty, Max commented on the situation, suggesting some changes in response to the dean's request that Max return to Cologne to help. First, Max expressed his disappointment about the development; he felt the faculty was gambling with the whole achievement:

> I thought we had created a new type of institute for Germany and Europe and hoped that would set an example. The main point for me was to set up a group of people within the university, that could compete with the best modern institutes in teaching molecular genetics and in the corresponding research. In addition to that I wanted to demonstrate ad oculos at a European university the polycephaly principle as it is known here. It is this principle and not the money that is the true secret weapon of the American universities.

Considering this "polyhead" principle nonnegotiable, Max suggested having five departments in the institute, with heads of departments—whether full professors or not—being codirectors of the institute. The five codirectors would decide among them on one chairman.

The decisive faculty meeting took place on July 29, 1964. Max attended the meeting, which opened at 3:00 P.M. At first, Max occupied his time writing a long letter home, until 5:00 P.M., when his suggestions became the next item on the agenda. After some discussion, three "historic decisions" passed: At 6:30 P.M. the new *Geschaftsordnung* (administrative configuration) was agreed to; at 6:50 P.M., the five chairs were renamed and reorganized *("Umbenenneung der Lehrstuhle")*; and at 7:15 P.M., all personnel decisions were settled. The meeting ended at 8:15 P.M. The institute was safe.

THE UNIVERSITY OF KONSTANZ

Several months before this historic faculty meeting in Cologne, the parliament of Baden-Württemberg had decided, in the spring of 1964, to build a new university in Konstanz. As founding president, they called upon Gerhard Hess, who had been president of the Deutsche Forschungsgemeinschaft when Straub applied for money for his Cologne institute. At a luncheon in 1956, Max had informed Hess about the importance of the new genetics. In early 1965, more than eight years later, Hess inquired if Max would be willing to help again. Individual scientists also solicited Max's help. Late in the same year, Klaus Bayreuther wrote to him: "Many colleagues . . . would like to come to Konstanz, if you would become the spiritual leader of its biology. They would love to take all the administrative work away from you ... they simply want a guarantee that this place has the right spirit from the very beginning. The rumor that you were asked to join the project has tremendously pushed the interest in Konstanz."

The "project" Bayreuther referred to was the plan to create in Konstanz something analogous to an American private university, with high admission standards. In Europe, private universities don't exist; all are state universities. The plan for a private university in Konstanz did not prove feasible, mainly because the accreditation of such an institution and the accreditation of the degrees awarded by it would have faced insurmountable difficulties. There was also a movement afoot at the same time to create at Konstanz an international European institute of technology on a large scale. The plans for that went quite far, though failing in the end. What was founded in Konstanz in 1966 was simply a modern state university.

Max replied to the 1965 requests. Remembering the difficulties in Cologne after his departure, he wrote to Hess on December 1, 1965: "It would be better in any case, if the faculty is created from the very beginning without a father figure." Max felt he was not needed as a professor in Konstanz, but only to advise on the initial selection of faculty and on the design of the biology department. He therefore joined the founding committee as a consultant for the Natural Sciences Faculty, thereby helping to construct a faculty focusing on molecular biology. He eliminated the classical disciplines of botany and zoology to streamline the department toward contemporary research. Max insisted on one chair for membrane biology, on the grounds that this was needed for "the next phase" in molecular biology.

The department that Max helped to create functioned reasonably well—and still does. In fact, it was not Max alone who made it work. When he decided not to join the faculty in Konstanz, quite a few scientists lost interest in the project. It required hard and skillful work of the first professors in Konstanz—foremost among them Peter Hemmerich and Horst Sund—to fill in with new names. They got Konstanz going and even convinced Max to accept a permanent guest professorship. Though he was to spend one term every three years in Konstanz, Max managed only one substantial visit to the new university that he considered "one of the very few successful universities founded in post-war Germany" (I). The Delbrücks came for the summer semester in 1969. These three months in Konstanz were to be Max's last lengthy stay in Germany.

1969—A DELBRÜCK YEAR

Max was back in Germany in December 1969. He stopped in Cologne on his way to Stockholm to receive the 1969 Nobel Prize in Physiology and Medicine. Max, Luria, and Hershey were jointly honored for their discoveries, which "first of all imply a deeper insight into the nature of viruses and virus diseases [Fig. 40]. Indirectly they also bring about an increased understanding of inheritance and of those mechanisms that control the development, growth and function of tissues and organisms. The work of the three . . . has great impact on biology in general," as the official announcement stated.

The first notification from Stockholm was received on October 16, 1969. After the phone rang at 5:25 A.M. breaking the news, Max was in for a busy day. He noted in his diary: "Incessant phone calls. Photographers at breakfast. . . . Press conference 10:30 A.M., lunch, brief siesta, celebration in lab. Tennis match (doubles) WON!"

Even before the pre-dawn phone call came, 1969 had been a great year for Max. In March the great *Phycomyces* review was published (D 74), and the summer was spent in Konstanz. From there, Max and Manny embarked with Jeanne Mammen on an adventurous trip to Morocco, fulfilling a more than thirty-year-old dream. It was

40. *The Phage Trio in Stockholm (1969). Left to right: Alfred Hershey, Salvadore Luria, and Max. Reprinted from Ivan Johansson, Meilensteine der Genetik. Hamburg and Berlin: Paul Parey, 1980.*

a risky undertaking from the very beginning. Jeanne Mammen was seventy-eight years old, and rather weak. The tour became dangerous when she caught a heavy cold and could only continue with great effort. When all three returned from the Sahara to Rabat on their tenth day in Morocco, Jeanne collapsed. She could no longer travel. Max and Manny had to leave her in a hospital where she recovered and was able to return to Germany ten days later.

In early October 1969, a week before the announcement of the Nobel Committee, Max shared the Louise-Gross-Horwitz Prize for genetics with Luria. During the ceremony at Columbia University, Max was rather talkative, pointing out that he owed his "life as a scientist to the fact that I did not remain in Germany during the Nazi days to participate . . . in the German resistance." Those who did stay, "and paid with their lives," Max called "prisoners of conscience." He donated his prize money to Amnesty International, "as a debt to all prisoners of conscience." Max had heard of the worldwide human rights organization through his sister, Emmi Bonhoeffer, who was active in one of the numerous German affiliates. In a statement, Max explained why it seemed to him fitting to handle his share of the money in that fashion: "If society expresses its debt to scientists by happenings like the present one, it seems to me that the scientist might as well express his debt to society, which permits him the pursuit of truth in a life exeptionally free of the constraints put upon most of its members."

Max was in a good mood in New York. He was also pleased when the news broke of his share of the Nobel Prize; as he cabled to Stockholm, he was "happy to share this prize with my friends," and he looked forward to a reunion in Sweden at the ceremony. Yet when the great day arrived—December 10, 1969—Max felt depressed, "like women after they gave birth to a child," as he wrote to Jeanne Mammen from Stockholm. "The main reason for my depression is my feeling guilty. All the time one is questioned about items one doesn't know anything about, though one should. These questions refer to a world outside the ivory tower which I used to ignore successfully."

Max concluded that Samuel Beckett had made the right decision. The author of *Waiting for Godot* had been awarded the 1969 Nobel Prize for Literature and had accepted the prize, but he did not attend the ceremonies, sending his publisher instead. Max had immensely enjoyed reading Beckett and looked forward to meeting him. After the festivities, Max went back to Beckett's writings, "reading Molloy, a few pages a day," as he noted in his diary. Molloy's motives did not seem to differ much from the those of a scientist (D 81).

Eventually, in 1976, Max did get a chance to meet the writer he admired. In that year, Max was invited to open an international symposium on the molecular basis of host-virus interaction at Hindu University in Benares. Since he was also invited to lecture at the Centennial of the Carlsberg Laboratory in Copenhagen, and had been expected for a long time to visit his friend and *Phycomyces* coworker—Tamotsu Ootaki—in Yamagata (Japan), Max decided to go around the world. On his way to Copenhagen in September 1976, he stopped in Berlin first. Beckett was also there, and a meeting was arranged. The two men met and walked along the Spree. Max asked Beckett about Molloy and the motivation of a scientist. Beckett answered that he had not thought about this angle. Max left, disappointed, for Copenhagen and points east.

A QUIET REUNION

After the Nobel celebrations in 1969 and before returning home, Max also traveled east to visit Timoféeff-Ressovsky in the USSR. In the fall of 1969 Max had decided that he would like to go to Russia to speak on behalf of his old friend, who had recently been removed from his institute and forced to retire on very small pay. On October 15, Max consulted with the Russian physicist Pjotr Kapitza, who advised such a visit. When Manny answered the phone early next morning and told Max about the Nobel Prize, his first reaction was: "Good, that will make it easier to go and see Timoféeff." A month later, about two weeks before Max's departure, arrangements were approved by the Russian Academy of Science.

Max hoped to obtain official rehabilitation for Timoféeff, whose "crime" was that he did not to return to Russia in 1933 when the Nazis came to power. He was accused of having collaborated with them. Timoféeff had already suffered punishment through a two-year confinement in a Siberian labor camp; he was in fact almost dead when the authorities decided to rescue him for his scientific abilities. In 1969, Max and Timoféeff did meet "unofficially" in the homes of supporting friends, who had been "officially" instructed the two should not be allowed to meet. Max was not, however, successful in arranging a meeting with those in power whose intervention he sought, and it remained unclear to him what effect his visit had on the future of his friend. Timoféeff's situation improved somewhat, but he remained a lonely and isolated man. He died in 1981, within a few weeks of Max's death.

THANK YOU

The year 1969 came to an end with Max's return to Caltech. In January 1970, Caltech gave a party to honor its Nobel Laureates and Max was asked to speak. After describing the festivities, he pointed out that he found one thing troubling (the following text is transcribed from a tape recording; the story that Max read is from Isak Dinesen's *Seven Gothic Tales*, 1934):

> I referred to the many parties that were in Stockholm, and in this connection I found one thing troubling. While the aspect of publicity and randomness were obvious to all of us, it was also clear that the Swedes themselves took enormous pride and pleasure in their parties. And they were wonderful parties, just because our hosts took so much pleasure in them. That was one of the great surprises. Even the Royal Family obviously did not consider this a chore, but their big occasion.
>
> So what do you do about that? I found [a] quote that I want to read, in a story where a young Italian girl and a Danish nobleman are talking about love and about parties. The girl says: I suppose that even in your country you have parties, balls, and conversazione? (This is taking place in Italy.)
>
> "Yes," he said, "we have those."
>
> "Then you will know," she went on slowly, "that the part of a guest is different from that of a host or hostess, and that people do not want or expect the same things in the two different capacities."
>
> "I think you are right," said Count Augustus.
>
> "Now, God," she said, "when he created Adam and Eve, arranged it so that man takes in these matters (in the matters of love) the part of a guest, and woman that of a hostess. Therefore man takes love lightly, for the honor and dignity of his house is not involved therein. And you can also surely be a guest to many people to whom you never want to be a host. Now tell me, Count, what does a guest want?"
>
> "I believe," said Augustus when he had thought for a moment, "that if we do, as I think we ought to here, leave out the crude guest, who comes to be regaled, takes what he can get, and goes away, a guest wants first of all to be diverted, to get out of his daily monotony or worry (and we certainly did in Stockholm). Secondly, the decent guest wants to

shine, to expand himself, and to impress his own personality upon his surroundings. And thirdly, perhaps, he wants to find some justification for his existence altogether. But since you put it so charmingly, signorina, please tell me now: What does a hostess want?"

"The hostess," said the young lady, "wants to be thanked."

The Stockholm festivities, at which my fellow laureates and I were entertained with such incomparable grace and splendor, left one thing wanting, which I found disturbing. In some cases it was difficult to identify the hostess to whom I would wish to express my thanks.

Now, this dinner tonight is the last of the parties that are going to be given in this connection. And, again, I find it difficult to define a hostess. So, whom should I thank? It would perhaps seem trivial to say you thank your wife for having been your hostess for the last 30 years, or all women in your life, starting with your mother. But I guess that is what it must be. So, in this sense, I'll conclude.

It might be more appropriate, though, to conclude the story of Max's spectacular year 1969 with the Japanese poem that he read at the end of the press conference in October and that he mailed (handwritten) to American friends who congratulated him on the award:

> The temple bell echoes the impermanence of all things
> The colors of the flowers testify to the truth
> That those who flourish must decay
> Pride lasts but a little while like a dream in a spring night
> Before long the mighty are cast down
> And are as dust before the wind.

Mind from Matter?

"THE ARROW OF TIME"

At the time Max entered the world of science in Germany, Robert A. Millikan had just founded the California Institute of Technology in Pasadena. The famous physicist designed a seal for that occasion, later explaining the reasons for his choice in a memo in 1925:

> The significance of the California Institute is embraced in its seal which represents a man in middle life running through clouds passing on a torch to a youth of twenty who overtakes him. The first concern of the Institute has been to have the torch passed on—the torch of human progress which is lighted at the flame of science. The second concern is to develop and inspire the youth who has the capacity to carry on that torch to the generations which are to come.

As a motto, Millikan chose a verse from the Gospel according to St. John (Chapter 8, Verse 32): "The truth shall set you free." As Max understood it, Millikan wanted to symbolize scientific truth and the progress it brings in the direction of enlightenment, liberation from superstition, and ultimately a better, more rational society. Although Millikan himself was a religious man, reconciling science and religion, many intellectuals of his generation believed that science would outpace and largely oust all other intellectual and spiritual endeavors. They were convinced that science would lead to a better world; many even thought it would displace religion by the end of the century.

More than half a century later, however, it had become obvious that the scientific culture had in no way eliminated the strength and the intensity of human religious needs or religious fervor. Max stressed this when in June 1978 he delivered the commencement address at Caltech, entitled "The Arrow of Time" (D 103). He noted a radical change from the trust in science characteristic of his student days. While he agreed that science constitutes the greatest intellectual advancement of the twentieth century, science is—in Max's view—intrinsically incapable of coping with the recurrent questions of "death, love, moral decisions, greed, anger, aggression." These are, however, the factors that determine man's values. They constitute the greatest forces "that shape man's destiny."

Science thus is immensely powerful and at the same time severely limited. Max felt that this human enterprise is doomed to be mistrusted by both the public and its own best students if it remains self-serving, if it continues in the blind faith that what is good for science is good for mankind. Max, though, felt that it is not the individual scientist who should set the necessary limits: "I think the scientist, insofar as he is a scientist, has to do what he did before. Scientific *institutions*, like Caltech, will have to become more involved in value questions" (I).

In a commencement speech given two years earlier, Max had explained what in his very personal case it meant to be a scientist. On May 26, 1976, at the California State University at Long Beach, he described his research on the sex life of his "little creature" *Phycomyces*, by both asking and answering the question, "Why study it, though?" He discounted potential applications:

Nothing could be further from my mind. It just happens to be a piece of research I am hooked on like an addict, or a nut who likes to solve puzzles. There is great happiness in doing research, coming in in the morning, burning to see what answer Nature has given to your experiment. If you are lucky and persistent you may make fantastic discoveries, totally surprising ones like the discovery of America and carrying the same overwhelming sense of hard and new truth.

This speech was entitled "The Pig and I." In it, Max offered pigs as the appropriate companion to live with in the future. Since a pig is down-to-earth, has its nose to the ground, does not look up to anybody nor down on anybody, he advised the students to hold on to the pig if they wanted to keep their "sanity in the difficult years ahead."

The unusual title Max chose for this, his first commencement speech, has its history in Delbrück family antics. The invitation came a day after Tobi Delbrück's sixteenth birthday. On that occasion, the Delbrücks gave a party highlighted by an extemporaneous puppet show. Ludina Delbrück, then fourteen, took the main part. Since she was fond of a little stuffed pig named Wilbur, the show was called *The Pig and I*. The plot was as follows: A professor—Max—was looking for someone to be his friend. A lion, a mouse, a dog, and a cat competed for the job. Yet the professor picked a pig, since "dogs look up to us; cats look down on us; pigs is equal." Max's characteristic combination of a spirit of humanity and a light sense of humor, displayed so vividly in family celebrations, became part of his formal appearance at a public event.

SCIENCE AND SOCIETY

After receiving the Nobel Prize, Max had to field much public attention which he considered "stupid." Among other things, he was asked to donate his sperm; representatives of several sperm banks kept pushing him. Max eventually turned all requests down by pointing out that he insisted on personal delivery.

In the early seventies, he was also asked what he thought about practically everything, including the relation between science and society (D 80). Max usually declined to comment. When, however, the science department of a Moscow newspaper, the *Literary Gazette*, sent him a list of twelve questions in November 1971, he did answer some of them, omitting answers to questions 2, 3, 6, 7, 9, and 12 (see the footnote on p. 282). The parts of his answers given in square brackets were omitted from the Russian publication.

(1) To what do you attribute the fact that in the last decade science has taken so important a place in the life of humanity? Will it keep this place in future?

Science started thousands of years ago. It became a partially organized enterprise of a happy few five hundred years ago. During the last decades it has become the greatest power on earth. Its developments might be likened to that of the Homeric legend of Tithonus and Aurora. The goddess Aurora (humanity) fell in love with a beautiful youth, Tithonus (science). They asked Zeus for the favor of giving Tithonus immortality. The request was granted. In due course an embarrassing situation developed. They realized that in asking for the immortality of Tithonus they had failed to include a request for eternal youth. While Aurora remained young and beautiful, Tithonus became old and shriveled. They requested Zeus to permit Tithonus to die, but it turned out, by some principles of irreversibility, that the "Gods themselves cannot recall their gifts" (Tennyson). Eventually a compromise was arranged according to which Tithonus was converted into a cricket. He was put into a box and permitted to chatter incessantly.

I think that the immortality of science, too, is irreversible. The problem is that of arranging a compromise analogous to the one which liberated Aurora from her embarrassing companion.

(4) Could, in your opinion, rapid development of science lead to some undesirable consequences?

The rapid development of science not only could lead to some undesirable consequences but has already done so. Science is potentially exceedingly harmful. The menace of blowing ourselves up by atom bombs or of doing ourselves in by chemical or biological warfare or by population explosions is certainly with us. In addition there are endless environmental problems such as lead poisoning, mercury poisoning, smog, waste disposal, and many others. By comparison with the group mentioned these may seem trivial questions, like housekeeping. They can be solved by appropriate legislation.

Perhaps the greatest threat to the future of humanity as well as its greatest hope lies in genetic engineering.

(5) Do you think that some fields of scientific research could be made taboo from the moral point of view? If so, which? Why?

I do not think that some fields of scientific research could be made taboo from the moral point of view. I do not think that there exists a possible analog to the Hippocratic oath which would ask all scientists to use their expertise and way of thinking to guard against the bad effects of science on society.

The reason that I think such a regulation is impossible is the following. The original Hippocratic oath says you should keep the patient alive under all circumstances; [also that you shouldn't be bribed, shouldn't give poison, should honor your teachers, and things like that, but essentially to keep the patient alive.] That is a reasonably well defined goal since keeping the patient alive is biologically unambiguous. But to use science for the good of society is not so well defined. Therefore I think that such an oath could never be written. Consider Einstein in 1905 concerning himself with a deep problem of the concepts of space and time in order to analyze some subtle paradoxes regarding the electrodynamics of moving bodies. At the time he did this work nothing could have seemed more impersonal, impractical, and more remote from any social implications. As is well known, these speculations eventually and inevitably led to the atomic arms race. Would anybody therefore wish to stop the future Einstein from thinking, or could they?

(8) Does it not seem to you that after a period of extreme popularity of the exact sciences among youth a cooling off may set in? (One of the

possible symptoms of this may be a decrease in the number of competitors for entree to educational institutions where science is taught.)

[I infer from your question that in the Soviet Union a decrease has occurred in the number of competitors for entry to educational institutions where science is taught.] I am not familiar with the corresponding statistics relating to this country, but it is certainly true that among the science students the interest in the social implications of science has vastly increased during the last few years, and correlated with that a great diversity of attempts to find new forms for personal experience and interpersonal relations. A greatly increased interest in literature and the arts is only one aspect of it.

[(10) Does it not seem to you that successes gained in the natural sciences give rise, in some scientists, to a somewhat disdainful attitude towards literature and the arts. If that is so, what should be done to counteract this?

I do not think that the somewhat disdainful attitude among some scientists towards literature and the arts is new. It is at least a hundred years old and on the whole on the wane. Science as well as traditional literature and art are on the defensive. Many of their protagonists feel that their successes have been hollow.]

(11) Does scientific research by itself foster high moral qualities in men?

Scientific research by itself fosters one high moral quality: that you should be reasonably honest. This quality is in fact displayed to a remarkable extent. Although many of the things that you read in scientific journals are wrong, one does assume automatically that the author at least believed that he was right. This belief has become so ingrained that if somebody deliberately sets out to cheat he can get away with it for years. There are a number of celebrated cases of cheating or of hoaxes that could illustrate this point. However our whole scientific discourse is based on the premise that everybody is at least trying to tell the truth.*

---

*Questions not answered by Max:
(2) What scientific development in recent time has made the greatest impression on you?
(3) What important scientific discoveries can be expected to be made in the foreseeable future? What would you like them to be?
(6) May there be any reason for stopping a successful line of research? If so, what are they?
(7) May keen public interest hinder the development of science?
(9) Can literature and the arts exert an influence on the development of scientific thought?
(12) Would you like your son or daughter to become a scientist? If so, in what field? And why?

It was typical for Max not to answer the twelfth question that inquired about his family. Even to his kids themselves, his advice was rather unconventional. One day, his daughter Nicola came to him for help: she couldn't make up her mind whether to study biology or dancing. Max recommended dancing as long as she was still young; later there would be still plenty of time left for biology. Nicola today teaches dancing.

Max responded to the Russian questions in the early seventies. The threat posed by genetic engineering became a much more urgent problem in the late seventies, when gene technology appeared on the horizon. Max was frequently asked to state his opinion on the issue; in response to one such request, a question the Delbrücks' friend Muriel Heineman posed during a telephone conversation, he replied in a peculiar way by writing an amusing poem, a "Valentine for N.I.H. [National Institutes of Health]." Max distributed the poem to others besides Muriel; eventually it was sent to *Trends in Biochemical Science,* where it was published. Max was pleased to see the Valentine in print, for it reflected his true opinion:

A blessing to us all is DNA
A child can understand its helix way.
Not only that, it now brings folks together
Who never thought themselves birds of a feather.

Friends separated over time and space
Find now an unsuspected common base.
From Muriel a valentine I got:
"Please tell me in a single shot
What all this gene-stuff is about.
Why all this anguish and this doubt?"

Here comes my answer, off the cuff.
Don't hesitate to phone, if not enough:

"Genetic Engineering is the game
That's older than the Bible, but its name
Ten thousand years ago was breeding.
It mixed the genes by sex and feeding.
The dog, the bean, the corn, the cat, the wheat
Were monsters made by man's Promethean heat.
All cultures grew on this tremendous theme:
"Manipulate the species that surround your deme,
Cross them, produce all kinds of puppies,
Keep what you want and feed the rest to guppies.

We have no record in our books
Of B.C. breeders tried as crooks,
That fears and anguish made the round
About some monstrous breed of hound,
One to devour the human race,
To kill us all without a trace."

"So what has changed since those old days,
What's new in this year's clever ways?"

"We now use chemistry to shuffle genes,
Use plasmids to move man's to beans,
Or rat's to microbes, flies' to fleas,
Or yeast's to coli, bee's to peas.

All this is based on Watson-Crick's
Phantastic Double Helix, plus some tricks
That others added to this play
And add still more from day to day."

"So why the worry, the commotion,
The horror stories, fearsome notions,
The panels, speeches, public hearing,
The guidelines, legislations, steerings,
A giant bureaucratic mire
To keep in check Prometheus' fire?"

"The answer in my humble view
Lies not with genes but lies with you
With fears that Science, by some foul play,
Will do us in one night or day.
I grant you that may well be true
But not by crossing man with flu!"

A STUDENT OF EVOLUTION

By the time this Valentine was published, Max had finished his very
last course on "Selected Topics in Biophysics." In 1975, Max had
devoted the course to the issue of "evolutionary epistemology." He
entitled his class "Truth and Reality in the Natural Sciences," and
epitomized it as an "investigation into human cognitive capabili-
ties, as expressed in the various sciences." In 1978, Max repeated
this course as a Professor Emeritus; a student, Judy Greengard,
taped the lectures and produced a transcript which Max edited
"mildly," giving it the title "Mind from Matter?"

In this rough form Max sent the text to a publisher, inquiring if
the manuscript would be of any interest. He offered to try to get it
ready for print in about a year's time, by the end of 1979. He also

considered accepting an invitation from Mount Holyoke to repeat the lectures, intending to use the opportunity to rework the material as a book. Until the autobiography, this was the only book Max ever contemplated writing. His illness intervened to prevent him from working on this project any further. The editing was taken up, however, by some of his friends, foremost among them Gunther Stent. *Mind from Matter?* was published in 1985.

In these lectures, Max described what he was learning as a "student of evolution," as he wrote Jeanne Mammen shortly before she died in 1975. The manuscript sketches his global philosophical vision. He did publish two highly condensed versions of the course (D 101, D 102), both of which are quite difficult to read.

A curious student, Max posed three questions:

(1) How can we construct a theory of a universe without life, and therefore without mind, and then expect life and mind to evolve somehow from this lifeless and mindless beginning?
(2) How can we conceive of the evolution of organisms with mind strictly as an adaptive response to selective pressures favoring specimens able to cope with life in the cave, and then expect that this mind is capable of elaborating the most profound insights into mathematics, cosmology, matter, and the general organization of life and mind itself?
(3) Indeed, does it even make sense to posit that the capacity to know truth can arive from dead matter? (D 111, p. 22).

Max began his exploration by tracing the origins of the universe and of living matter in particular, moving to an examination of the evolution of the human brain and of perception. Initially, he assumed a philosophical stance of naive realism for the intuitive power it allows; in the second half of the course, he showed how "modern scientific knowledge forces revisions of the naive view of reality with which we set out" (D 111, p. 22). Early in his discussion, Max pointed to research, both neural and psychological, showing that vision involves a high degree of "abstracting and filtering of sensory information at preconscious levels" (p. 273). Many steps intervene in the brain between the primary sense impression and the visual experience of it; consciousness has no access to raw, unprocessed sensory data. Moreover, such an abstracting process is not simply acquired by each individual through experience. On the contrary, as Immanuel Kant suggested two hundred years ago, the organization of the human brain "endow[s] us with forms of cognition that . . . appeared to be a priori, that is, prior to any experience" (p. 273). We employ such built-in, a priori categories to classify and mediate our perceptions.

To answer how our a priori categories "happen to fit the real world so well," Max pointed to the evolutionary explanation provided by Konrad Lorenz: what is built into the individual, and thus is prior to the individual's experience, has already been learned by his species. Adaptation to the real world has preserved the most successful circuitry for perceiving reality. However, our cognitive categories and associated circuitry are not fully developed at birth. Instead, they unfold through a "dialectic interaction between the developing nervous system and the real world" (p. 274). To show the active nature of the human mind in processing sensory input, Max turned to the work of Jean Piaget, who in the 1920s opened what Max described as "a gold mine for epistemological exploration, one that was overlooked for millennia by philosophers" (p. 125). Piaget's observations of the development of cognitive functions in individual children led to the conclusion that the categories of cognition inherent in the mind develop during childhood in an invariant and orderly sequence of separate stages.

Thus, Max pointed out, the human mind's built-in cognitive categories result from evolutionary adaptations to life in the real world, which Max called the "world of middle dimensions, the world of our direct experience." Such categories would be expected to require modification when our activities move beyond the middle dimensions, as they do in modern physics, astronomy, or biology, where we study "the very small and the very brief," or "the very large and the very long." In such studies we encounter difficulties, for as we use middle-dimension categories to go beyond this realm of direct experience, we are bound to end up in paradoxical situations.

In order to show this in detail Max focused on several sciences, beginning with mathematics. Here he described the paradoxes arising in basic set theory and in David Hilbert's attempt in the early part of the twentieth century to remove paradox, inconsistency, and incompleteness from the realm of mathematics. As a student in Göttingen, Max had participated in Hilbert's seminar on "The Structure of Matter." Shortly after, in the 1930s, Kurt Gödel "pulled the rug out from under these efforts" by showing that Hilbert's goal is unattainable, for in every complex mathematical system one can find inconsistent and unprovable propositions.

With his presentation of Gödel's proof and its implication for him that mathematics is not independent of the human mind, Max concluded that our conceptual categories function well in the realm of direct experience. But we will encounter paradoxes in trying to apply the category "number" to things that transcend our experience, as infinite sets do.

Max then turned to the physical sciences, showing that by 1900 classical physics had constructed a "satisfactory, wholly deterministic picture of the world, built on seemingly self-consistent notions" (p. 199). But the picture was even then not wholly satisfactory, for Lord Kelvin in 1900 pointed to some "clouds on the horizon of classical physics," areas in which predictions did not match with measurement. In particular, predictions of statistical mechanics did not conform with the behavior of electromagnetic waves. The resolution of the discrepancy led to the rise of quantum mechanics, leading eventually to the notion of complementarity as introduced by Bohr.

Max termed quantum behavior a "conspiracy of nature that prevents us from attaining a fully deterministic description of physical phenomena" (p. 225). In every experimental observation of quantum phenomena, the observer is engaged in active, subjective decisions that limit what he can observe. At the quantum level, measuring devices and experimental set-ups form a part of the physical reality and disturb the system. Furthermore, quantum theory forces us to acknowledge other, counterintuitive characteristics of reality besides the fact that quantum reality is not fully knowable: quantum particles have no definite trajectories but only probabilistic ones; they have no distinct identities; and they are not conserved. Moreover, even in pre-quantum physics, the laws cannot be separated from the personal experiences of observers. There is no reality independent of our experience of it; there is no clean-cut split between mind and matter. Yet realism requires such a split. All in all, Max concluded that the insights of quantum mechanics lead us to dismiss even sophisticated versions of realism.

Max next confronted the paradox of the evolution of mind: if mind has developed cognitive categories to cope with the directly experienced physical world, how can it so successfully deal with a world of alternative realities for which it was not selected? Why is the world of quantum mechanics at all comprehensible to us? His answer was that the world of complementarity resembles an infant's evanescent world, where objects have no permanence and are not conserved. We do not lose the cognitive antecedents required to deal with such a world, even while we remain saddled with the concrete mental operations refined through evolution.

Finally, he addressed the widespread belief that mind transcends physical phenomena, that mind transcends matter. The human capacity for language is the biggest stumbling block for any view of mind as material phenomena; no machine thus far has semantic competence. Yet, as Max pointed out, organic evolution has had quite a headstart on any computer designers in creating a system capable of semantic decoding. The absurdity we feel in the question "Mind from Matter?" may prove as mutable as the absurdity of the anti-intuitive concepts of quantum theory.

Max's presentation in these lectures left many questions unanswered, including the question of the title: "Mind from Matter?" He encouraged his students to answer these questions, even to give snap answers which he considered better than no answers. He asked the students to help bring the questions into sharper focus and to spend the rest of their lives trying to account for them.

SCIENCE AND POETRY

While preparing "Mind from Matter?" Max became interested in the structure of languages, partially because language is the property that now distinguishes man from machines, but also because science and poetry have language in common. Max discussed the relation between these two intellectual ventures and their language use in a number of letters he exchanged with the Muriel for whom he composed the Valentine. Shortly after Max finished his epistemological lectures, he wrote (1978):

> One of the bizarre aspects of *science,* is of course, that it prides itself that it uses language accurately, and, by implication, that everybody else is wobbly. How come, then, that poets can use the same words, use them in a totally different way, and yet, in some cases, be fabulously precise writers: alter a single word and the essence is gone. To what, in us, does the poet speak? I am puzzled by the dichotomy, in us, between the Universe of Discourse of Science, and one or more other Universes of Discourses (of Faith, even!) that use the same words.

Indeed, Max's last thoughts centered around the question of what it means to use words with precision in poetry. He particularly admired the poetry of T. S Eliot *(The Four Quartets)* and Rainer

Maria Rilke. The *Duino Elegies* fascinated him all his life. In August 1980, Max answered a request from the Poetry Center in New York that he lecture on the relation between science and poetry: "The Duino Elegies have been much on my mind during the last fifty years and I was wondering whether I could verbalize in some form or other the enduring and monumental shock they produce."

After some thought, Max began to prepare the talk, entitled "Rilke's Eighth Duino Elegy and the Unique Position of Man." Scheduled for the fall of 1980, it had to be canceled due to Max's health problems. Only a partial manuscript exists. In it, Max emphasized that the use of language by poets and by scientists is of an entirely different nature. In poetry, words have their meaning in the context of the whole language, whereas in science, language represents an ultimate in restriction. Each word is to be understood independent of any context but the scientific one. Nevertheless, as Max noted, "there are striking similarities in the intellectual operation of the poet and the scientist, especially in the relation between the conscious labor and the unconscious labor."

Rilke's *Duino Elegies* were written in two major installments in 1912 and in 1922. The first part came to Rilke as a voice speaking to him "out of the wind," or out of the unconscious mind. Scientists report similar experiences wherein unconscious thoughts break into consciousness (Hadamard, 1949). To Max, the elegies demonstrated yet another striking similarity between the creative processes in poetry and in science: "the essential involvement of intellectual shocks." Max described this aspect in some detail:

> In science we see it where the creative mind is willing to accept an irrational and highly anti-intuitive situation, like Einstein in 1905 postulating that light has a particulate quality, discontinuous quanta, in spite of everything that was then known, proving its wave nature and continuous distribution; and in the same year, relativity theory, starting from the contradictory assumptions that the speed of light is the same for all inertial systems; or Niels Bohr in 1913 introducing the totally irrational assumption of electrons jumping in a nonmechanical way from one orbit to another. These ideas represent *intellectual shocks,* not linguistic ones. The shocking scientific papers are not written in poetic language. The last scientist who published his excellent science in verse was Lucretius. He explicitly says that he put his ideas in verse to make them more agreeable reading, not to bring out deep ideas requiring poetry for their expression.
>
> In poetry the shock is there, too, but it is of a different nature. Every one of the Duino Elegies starts with a shock. Let us say a few words

about the whole set and then examine the first lines of the first elegy.

There are 10 elegies, each about 90 lines; too much to read in one sitting, perhaps too much to take in a lifetime. They are unrhymed and written free in rhythm. They all concern man's position in the universe and thus express Rilke's highly personal metaphysical situation. Some of the elegies conform to the conventional definition of elegy, especially the first ones, being laments. Some are descriptive; some meditate; the 9th is affirmative; the last one is jubilant in an ecstatic fashion. The reader gets a feeling of ritual song and dance in praise of the god Dionysos although the substance is totally different.

There have been about ten translations into English starting with the excellent one by Leishman and Spender in 1939, a German scholar and a poet collaborating—a masterpiece. The later translations, though probably slightly improved in scholarship, seem to suffer from the fact that each translator wants to be "independent" of his predecessors, i.e. different. It is my impression that as a result of this negative selection the translation gets gradually stiffer and thereby poorer, but the impression may well be too sweeping.

I would like to give you a feeling for the shocks with which the elegies open by examining just the first two lines of the first elegy. They will suffice to illustrate the shock in symbol, in vocabulary, in rhythm.

"Wer, wenn ich schriee, horte mich denn aus der Engel Ordnungen und gesetzt selbst es trate einer hervor . . ."

"Who, if I screamed, would hear me of the angelic orders
And, suppose even, one of them stepped forth . . ."

What angels? They are thrown at you without preparation and they recur throughout the ten elegies. They are certainly not Christmas tree angels and not archangels. Rilke hinted in a well-known letter to his puzzled Polish translator that they are more related to the Islamic angels. Every commentator has since then echoed this rather evasive hint, or has constructed his own angels. They are Rilke's angels, at this point undefined quantities, the way the mathematicians start out their systems. The next lines tell us that they are remote, eternal, self-sufficient, radiations reflecting back onto themselves, inaccessible to us, yet terrifying.

To illustrate the shock value of words let us take the word "schriee," the subjunctive of "schreie," an extremely strong word which I have translated "screamed." No other translator has done that. They translate with "cried," "called," etc., but "screamed" would seem to be the only one that implies the extreme anguish of "schriee."

Thirdly as to rhythm, the second line has the phrase "gesetzt selbst" (suppose even); a shocking apposition of two strongly accented syllables with massed sibilants and explosive consonants which give the feeling of a diver jumping from the diving board, an effect that could hardly be reproduced in English, and none of the translators has attempted to do so. The phrase itself seems dry and harmless enough. It could occur in any mathematical or legal text but Rilke's use of it in a poem is, I believe, unique.

I now want to turn to the 8th elegy which concerns the difference between animals and man. I would like to relate Rilke's intuitive and speculative assessment and description of the difference with a more scientific approach and see how the two compare.

The oldest statement about the difference between animals and men, by a poet at that, is by Aeschylus in "Prometheus Bound." Prometheus, just having been nailed to the rock by Zeus's helpers, lists in a magnificent speech to a chorus of sea nymphs who listen sympathetically all the good things he. . . .

Unfortunately, Max never completed the paper.

THE DEATH OF A SCIENTIST

In 1978, two days before Max was to have checked into Huntington Memorial Hospital in Pasadena for heart-bypass surgery, a multiple myeloma was discovered. This cancer caused his death in March 1981.

Since 1963, on the visit to Germany, Max had suffered from occasional chest pains. Angina pectoris was diagnosed by the German doctors but there was not really much one could do in the 1960s; bypass operations were not available then. Max decided on two tactics. He would not mention his pains to the family, and he would discuss all possible treatments with his Pasadena physician, Don Moore, in whom he had great trust. Moore advised Max that the standard recommendation for the duration of such chest pains was to sit down and rest. However, a new suggestion had just been described in the medical literature: to walk away the pain. While Moore personally sympathized with the new treatment, he did make clear that there was no consensus. Max favored the new approach (which later became accepted as standard treatment). Not only did this give him a chance actively to fight his pain but, in addition, the theory appealed to him. Walking was supposed to produce sufficient vasal dilation to overcome obstacles in the vessels.

Indeed, this approach did work for more than a decade, for Max suffered no chest pains in the early 1970s. In 1978, however, he was hit again by angina, and medical investigations showed advanced coronary disease.

In the seventies other medical problems plagued Max. Even in 1975 he feared he would soon "join the Club Downward and Out, to which we all belong perhaps from the day of our birth." He wrote this in a letter six years before he died, after noticing during a neurobiochemistry seminar that his vision was deteriorating. In the seminar room, he checked that his glasses had no spots; he then began to worry that the trouble might be originating in the brain. By closing his right and left eye alternately, he quickly noticed that the right half of the visual field was missing. At the end of the talk, Max discussed his symptoms with the neurobiologists present. They concluded that something must be wrong peripheral to the optical chiasma, probably in the left eye. An eye examination identified the problem as a retinal detachment demanding immediate surgery.

Max recovered from the eye problem and felt strong during 1976 and 1977, but in late 1977 began to experience occasional heart pains while bicycling to the laboratory. A bypass operation was scheduled for April 1978. Two days before the planned date, routine X-rays showed that his ribs were being chewed away from the inside. The physicians called off the bypass, telling him, "Forget about your heart, it is the least of your troubles." A bone scan the next day revealed that Max was suffering from multiple myeloma. This was May 2. He went to the next scheduled meeting of his research group, sitting through all the presentations. Then Max informed them about his disease. Three days later, he invited them to a seminar on myeloma.

While first discussing the cancer with Moore, Max wondered if there might be any connection between the two diseases plaguing him. Max reasoned that his blood might be clogged by the proteins that the cancerous cells were pouring into the bloodstream. That way, he could at least understand where the angina originated. He challenged his physician as to why he hadn't thought of that as a cause.

Chemotherapy and radiation treatments arrested and relieved the symptoms, and Max was able to return—for some time—to nearly normal activity. He resumed travel and camping trips, and kept an eye on the experiments designed to identify the photoreceptor in *Phycomyces* (Figs. 41a, 41b). In a sense, Max even enjoyed being a patient and seemed to love malingering. He received a lot

41a. *A Delbrück family backpacking trip to Devil's Canyon in the San Gabriel Mountains above Pasadena (May 1979). Left to right: Rick Salmon and his wife, Nicola Delbrück, Ludina Delbrück, Manny Delbrück, Jonathan Delbrück and his former wife Suzanne, Tobi Delbrück, and Max. Courtesy of Manny Delbrück.*

41b. *Max and Manny Delbrück at home in Pasadena (July 1980). Photo by Joanne Kronick Chris © 1980.*

of attention and a lot of visitors. Even in his illness, Max derived pleasure from having people around. He also advanced medical understanding of the treatment of cancer by allowing himself to be given the experimental drug interferon. Max was "willing to be a guinea pig," involved in the intellectual challenge of understanding the causes and treatment of the disease.

Gradually, however, the treatments as well as the disease took their toll. Max declined slowly, giving up hope of a recovery. He discontinued a special diary he had been keeping. He had opened this diary in the fall of 1978 when he was confined to a wheelchair for weeks, as follows:

> Being bored with myself, not having a major job at hand. Feeling dense in the head, for weeks already. . . . This diary is an attempt at some continuity of effort. Perhaps make it impersonal like Kierkegaard's . . . and give expression to the externality of the body to the self. . . .

Max had given his diary the German title *Die Heimreise (Coming Home)*. He quoted the German poet Novalis: "Wohin gehen wir denn? Immer nach Hause." (Where are we going? Always home.) When he was dying, Max felt that he was coming home.

He was aware of the fact that "one of the aspects that distinguishes humans from other creatures is that we can contemplate our own death" (D 89). He also knew that a limit exists to such thoughts; rationality is unable to cope with the prospect of death. Any attempt at rationally reconciling the complementary notions of life and death "can only lead to garbled nonsense about a God with spatial, temporal or material attributes, yet strongly interacting with our will, our thoughts and our action. This kind of compromise satisfies neither the intellect nor the instinct."

Max had said this in 1966 in a seven-minute speech he gave as part of a symposium on "Science and the Christian Faith" at Caltech. On that occasion he had pointed out the difficulties in combining a scientist's attitude as a practitioner of science with his "instinctual needs of Faith." Max exemplified the problem a scientist can run into when asked to express his faith by quoting Lucky's speech in Beckett's *Waiting for Godot*, when Lucky is ordered by his master to "Think, pig."

> Given the existence as uttered forth in the public
> works of Puncher and Wattman of a personal
> God quaquaquaqua white beard
> quaquaquaqua outside time without extension
> from whom the height of divine apathia divine
> athambia divine aphasia loves us dearly with

some exceptions for reasons unknown but time
will tell and suffers like the divine Miranda with
those who for reasons unknown but time will tell
are plunged in torment plunged in fire. . . . (Act One)

Max's life had been spent trying to solve "the riddle of life." He became aware that there are two ways to look at life. To a scientist, life is as precious, as wonderful, as mysterious as it is to any other man. But to the scientist it is also, quite convincingly, ephemeral, local, transient, an infinitesimal episode in the universe. In discussing this transient aspect of life, Max would quote Beckett's words, attributing them to *Molloy:*

> We give birth astride a grave, the light gleams an instant, then there is night once more.*

Max repeated this line over and over. His fascination was perhaps intensified because it equates light with life, the pairing to which Max had devoted so much of his scientific career.

Max had wanted to stay home with his family as long as possible; he had to be hospitalized on the last night, when his pains became unbearable. He died in the small hours of March 10, 1981, with Manny at his side.

*This is Max's version of the quotation, as it appears in his letters and notes. The correct quote, from *Waiting for Godot,* is: "They give birth astride of a grave, the light gleams an instant, then it's night once more."

Epilogue
The Inner Unity

In April of 1981, a memorial service took place at Caltech to honor Max. At his request, his friends and colleagues heard his favorite music (Bach's Cantata #106). They also heard Max on tape, and viewed slides depicting Max enjoying himself with family, fellow scientists, and students. All left reminded of how much fun Max had been, of how strong a personality he was, and of how deeply he had affected those with whom he came in contact.

What they did not hear is what this book has shown: the attraction to complementarity that motivated Max throughout his career. As excited as he was by the advances in understanding resulting from his work and in biology in general, he remained disappointed that none of the advances yielded the paradox he sought. Max's faith in the paradox steadily guided many of his career choices. It is one type of "inner unity" Max set out to find in his career when he undertook to write an autobiography.

A second "inner unity" might be that Max liked to explore unknown territory, to open up new fields, to hike and camp in undeveloped desert. The phage exploration he led did yield discoveries. The *Phycomyces* enterprise, chosen for the simplicity of the organism, did not provide immediate breakthroughs, though it is providing insight into sensory processing as fruitfully as any organism. None yields simple answers. The closer biologists look, the more complex they find life processes to be. And despite their discoveries Max might still have responded to them, "I don't believe a word of it."

In writing about Max, one must wonder how he might have reacted to this discussion of his scientific career. Would he have placed the volume on a shelf in a bathroom, as he did the Goethe set, to read in quiet moments? Might he have given us a variant of the pronouncement so many others met with: "This was the worst seminar I ever heard"? Might he have been found reading this book, as he did many others, in the midst of a party at which the discussion bored him?

We might even picture Max, in shorts with no shirt, in the middle of a party in his living room, stretched out on the floor to ease his back, chuckling at some anecdotes we've presented or protesting details. Inevitably, he would have challenged us in spots with his characteristic charge: "I don't believe a word of it." And we would have marshaled documents—his letters, and all the records of interviews and sources collected—to rise to the challenge of this most demanding and difficult, but brilliant and humane colleague: Max Delbrück.

Bibliography

Festschrift for Max Delbrück (cited as PATOOMB) (1966). G. Stent and J. D. Watson, eds., *Phage and the Origins of Molecular Biology*. Cold Spring Harbor Laboratory of Quantitative Biology, Cold Spring Harbor, N.Y.

Abir-Am, P. (1982). "The Discourse of Physical Power and Biological Knowledge in the 1930's: A Reappraisal of the Rockefeller Foundation's 'Policy' in Molecular Biology," *Soc. Studies of Science*, 12:341–382.

Adam, G. (1967). "Nervenerregung als kooperativer Kationenaustausch in einem zweidimensionalen Gitter," *Ber. Bunsenges. Phys. Chem.*, 71:829–831.

———. (1973). "Cooperative Transitions in Biological Membranes," in H. Haken, ed., *Synergetics*. Stuttgart: Teubner Verlag, pp. 220–231.

Allen, G. (1978). *Thomas Hunt Morgan: The Man and His Science*. Princeton, N.J.: Princeton University Press.

Anderson, T. F. (1949). "The Reactions of Bacterial Viruses with Their Host Cells," *Botan. Rev.*, 15:464–505.

——— (1966). "Electron Microscopy of Phages," in PATOOMB.

——— (1981). "Reflexions on Phage Genetics," *Ann. Rev. Genet.*, 15:405–417.

Armitage, P. (1952). "The Statistical Theory of Bacterial Populations Subject to Mutation," *J. Roy. Statist. Soc.*, B14:1–40.

Avery, O. T., MacLeod, C. M., and McCarthy, M. (1944). "Studies on the Chemical Nature of the Substance Inducing Transformation of Pneumococcal Types. Induction of Transformation by a Desoxyribonucleic Acid Fraction Isolated from Pneumococcus Type III," *J. Exptl. Med.*, 79:137–158.

Beadle, G. W., and Tatum, E. L. (1941a). "Genetic Control of Biochemical Reactions in Neurospora," *Proc. Natl. Acad. Sci. U.S.A.*, 27:499–506.

——— (1941b). "Genetic Control of Development and Differentiation," *Amer. Naturalist*, 75:107–116.

Beadle, G. W. (1966). "Biochemical Genetics: Some Recollections," in PATOOMB, 23–32.

Beese, W. (1980). "Die Arbeit von Max Delbrück und die Entstehung der Molekularbiologie," Dissertation, Humboldt-Universität, East-Berlin.

Benzer, S. (1952). "Resistance to Ultraviolet Light as an Index to the Reproduction of Bacteriophage," *J. Bacteriol.*, 63:59–72.

———— (1961). *Genetic Fine Structure.* New York: Harvey Lectures 56, Academic Press.

———— (1966). "Adventures in the rII Region," in PATOOMB, 157–165.

Bethe, H. A., and Rohrlich, F. (1952). "Small Angle Scattering of Light by a Coulomb Field," *Physical Review,* 86:10–16.

Bergman, K., Eslava, A. P., and Cerdá-Olmedo, E. (1973). "Mutants of *Phycomyces* with Abnormal Phototropism," *Mol. Gen. Genet.,* 123:1–16.

Björn, L. O. (1976). *Light and Life.* London: Hodder and Stoughton.

Boeck, J., Kaissling, K. E., and Schneider, D. (1965). "Insect Olfactory Receptors," *Cold Spring Harbor Symposium on Quantitative Biology,* Vol. XXX ("Sensory Receptors"), 263–280.

Böhme, H. (1966). *Deutschlands Weg zur Großmacht.* Cologne: Kiepenheuer und Witsch Verlag.

Bohr, N. (1928). "The Quantum Postulate and the Recent Development of Atomic Theory," *Nature,* 121:580.

———— (1931). *Atomtheorie und Naturbeschreibung.* Berlin: Julius Springer.

———— (1932). "Chemistry and the Quantum Theory of Atomic Constitution," *J. of the Chem. Soc.,* 349–384.

———— (1933). "Light and Life," *Nature,* 131:421 and 457; in German: "Licht und Leben," *Naturwissensch.,* 21:245.

———— (1963). *Atomic Physics and Human Knowledge.* New York: John Wiley & Sons; in German: *Atomphysik und menschliche Erkenntnis.* 2 vols., Braunschweig: Vieweg Verlag, 1962–64.

———— (1963). "Licht und Leben—noch einmal," *Naturwissensch.,* 50:-725.

Bose, S. N. (1924). "Planck's Law and the Hypothesis of Light Quanta," *Zeitsch. f. Physik,* 26:178.

Brenner, S. (1957). "On the Impossibility of All Overlapping Triplet Codes in Information Transfer from Nucleic Acids to Proteins," *Proc. Natl. Acad. Sci. U.S.A.,* 43:687–694.

Bresch, C., and Hausmann, R. (1970). *Klassische und molekulare Genetik.* Berlin: Julius Springer-Verlag.

Bridges, C. B. (1916). "Nondisjunction as a Proof of the Chromosome Theory of Heredity," *Genetics,* 1:1–52.

Bünning, E. (1935). "Sind Organismen mikrophysikalische Systeme?", *Erkenntnis,* 5:337–347.

Carlson, E. A. (1966). *The Gene: A Critical History.* Philadelphia: W. B. Saunders.

———— (1981). *Genes, Radiation, and Society, The Life and Work of H. J. Muller.* Ithaca, N.Y.: Cornell University Press.

Cerdá-Olmedo, E. (1977). "Behavioral Genetics of Phycomyces," *Ann. Rev. Microbiol.,* 31:535–547.

Chadwick, J. (1932). "Possible Existence of a Neutron," *Nature,* 129:312.

Clayton, R. (1973). "Primary Processes in Bacterial Photosynthesis," *Ann. Rev. Biophys. and Bioeng.,* 3:131–156.

———— (1970). *Light and Living Matter.* 2 vols., New York: McGraw-Hill.

Creighton, H. B. and McClintock (1931). "A Correlation of Cytological and Genetical Crossing-over in *Zea mays,*" *Proc. Natl. Acad. Sci. USA,* 17:492–497.

Crick, F. H. C., Griffith, J. S., and Orgel, L. E. (1957). "Codes Without Comma," *Proc. Natl. Acad. Sci. U.S.A.,* 43:416–421.

Courant, R., and Hilbert, D. (1928). *Methoden der Mathematischen Physik.* Berlin: Julius-Springer Verlag; in English (1966). *Methods of Mathematical Physics.* New York: Wiley.

Craig, G. A. (1943). "Delbrück: The Military Historian," in E. M. Earle, ed., *Makers of Modern Strategy, Military Thought from Machiavelli to Hitler.* Princeton, N.J.: Princeton University Press.

———— (1981). *Germany 1866–1945.* London: Oxford University Press.

de Vries, H. (1901). *Die Mutationes Theorie.* Leipzig: Veit Verlag.

Delbrück, B. (1904). *Einleitung in das Studium der indogermanischen Sprachen.* Leipzig: Breitkopf und Härtel.

Delbrück, H. (1900–36). *Geschichte der Kriegskunst im Rahmen der politischen Geschichte.* 7 vols., Berlin: G. Stilke Verlag.

———— (1918–19). *Krieg und Politik,* 3 vols. Berlin: O. Stellberg Verlag.

———— (1922). *Ludendorffs Selbstporträt.* Berlin: Verlag für Politik und Wirtschaft.

———— (1924–28). *Weltgeschichte.* Berlin: O. Stellberg Verlag.

Demerec, M. (1951). Foreword, *Cold Spring Harbor Symposium on Quantitative Biology,* XVI: p.v.

————, and Fano, U. (1945). "Bacteriophage-resistant Mutants in Escherichia coli," *Genetics,* 30:119–136.

Demerec, M., Adelberg, E. A., Clarck, A. J., and Hartmann, P. E. (1966). "A Proposal for a Uniform Nomenclature in Bacterial Genetics," *Genetics,* 54:61–76.

Dickson, H. (1932). "The Effects of X-rays, Ultraviolet Light, and Heat in Producing Saltants in Chaetomium cochliodes and Other Fungi," *Ann. Botany,* 46:389–405.

Dinesen, I. (1934). "The Roads Round Pisa," in *Seven Gothic Tales.* New York: Alfred A. Knopf.

Dirac, P. A. M. (1928). "The Quantum Theory of the Electron," *Proc. Royal Soc. Lon.,* A117:610–624.

———— (1931). "Quantised Singularities in the Electromagnetic Field," *Proc. Royal Soc. Lon.,* A 133:60–72.

Dirac, P. A. M. (1957). *The Fundamental Principles of Quantum Mechanics.* London: Oxford University Press; in German (1930). *Die Prinzipien der Quantenmechanik.* Leipzig: Hirzel Verlag.

———— (1978). "The Prediction of Antimatter," H. R. Crane Lecture, April 17, 1978, University of Michigan, Ann Arbor, Mich.

Doermann, A. H. (1948). "Lysis and lysis inhibition with Escherichia coli bacteriophage," *J. Bacteriol.,* 55:257–276.

———— (1952). "Intracellular Phage Growth as Studied by Premature Lysis," *Fed. Proc.,* 10:591–594.

Drieschner, M. (1981). "Einführung in die Naturphilosophie," Wissenschaftliche Buchgesellschaft, Darmstadt.

Dulbecco, R. (1949). "Reactivation of UV-inactivated Bacteriophage by Visible Light," *Nature,* 163:949–950.

———— (1966). "The Plaque Technique and the Development of Quantitative Animal Virology," in PATOOMB, 187–291.

Dunn, L. C. (1965). *A Short History of Genetics.* New York: McGraw-Hill.

Edgar, R. S., and Lielausis, I. (1964). "Temperature Sensitive Mutants of Bacteriophage T4D," *Genetics*, 49:649–662.

Edgar, R. S. (1966). "Conditional Lethals," in PATOOMB, 166–172.

Edidin, M. (1974). "Rotational and Translational Diffusion in Membranes," *Ann. Rev. Bioph. Bioeng.*, 3:179–201.

Einstein, A. (1905). "Über einen die Erzeugung und Verwandlung des Lichtes betreffenden heuristischen Gesichtspunkt," *Ann. Phys.*, 17:132.

Ellis, E. L. (1966). "Bacteriophage: One-Step Growth," in PATOOMB, 53–62.

Erikson, E. (1958). *Young Man Luther.* New York: W. W. Norton.

Eslava, A. P., and Cerdá-Olmedo, E. (1974). "Genetic Control of Phytoene Dehydrogenation in *Phycomyces,*" *Plant Sci. Letters*, 2:9–14.

Fechner, G. T. (1860). *Elemente der Psychophysik* (Leipzig). Amsterdam: Nachdruck (1964).

Feyerabend, P. (1968). "On a Recent Critique of Complementarity I," *Philos. of Science*, 35:309–331.

―――― (1969). "On a Recent Critique of Complementarity II," *Philos. of Science*, 36:82–105.

―――― (1975). *Against Method.* New York: New Left Books.

Feynman, R. P., et al. (1963). *The Feynman Lectures on Physics.* Cambridge, Mass.: Addison-Wesley.

Fisher, R. A. (1935). *The Genetical Theory of Natural Selection.* Oxford: Clarendon Press. (1958) new edn. New York: Dover Publications.

Fleming, D. (1969). "Emigré Physicists," in *The Intellectual Migration: Europe and America, 1930–1960.* Cambridge, Mass.: Harvard University Press.

Foster, K., and Lipson, E. D. (1973). "The Light Growth Response of *Phycomyces,*" *J. Gen. Physiol.*, 62:590–617.

Fox-Keller, E. (1983). *A Feeling for the Organism.* San Francisco: W. H. Freeman.

Fraenkel-Courat, H. (1956). "The Role of the Nucleic Acid in the Reconstitution of Active Tobacco Mosaic Virus," *J. Am. Chem. Soc.*, 78:882–883.

Friedrich-Freska, H. (1961). "Genetik und biochemische Genetik in den Instituten der Kaiser-Wilhelm-Gesellschaft und Max-Planck-Gesellschaft," *Naturwissensch.*, 48:10–22.

Frisch, O., and Meitner, L. (1939). "Products of the Fission of the Uranium Nucleus," *Nature*, 143:471; and "Disintegration of Uranium by Neutrons," *Nature*, 143:239.

Frisch, O. (1970). "Lise Meitner 1878–1968," *Biogr. Mem. of Fellows of The Roy. Soc.*, Vol 16 (November), pp. 405–420.

―――― (1979). *What Little I Remember.* Cambridge, Engl.: Cambridge University Press.

Fuerst, J. A. (1982). "The Role of Reductionism in the Development of Molecular Biology: Peripheral or Central?", *Soc. Stud. of Science*, 12:241–278.

Gaffron, H., and Wohl, K. (1936). "Zur Theorie der Assimilation," *Naturwissensch.*, 24:81–90.

Gamow, G. (1931). *Constitution of Atomic Nuclei and Radioactivity.* Oxford: Clarendon Press.

Gamow, G. (1954). "Possible Mathematical Relation Between DNA and Protein," *Biol. Medd. Dan. Vid. Selsk.*, 22, No. 2.

―――― (1960). *Physics. Foundations and Frontiers.* Englewood Cliffs, N.J.: Prentice-Hall.

———— (1966). *Thirty Years That Shook Physics*. New York: Anchor Books.

Gierer, A., and Schramm, V. (1958). "Infectivity of Ribonucleicacid from Tobacco Mosaic Virus," *Nature*, 177: 702–703.

Golomb, S. (1982). "Max Delbrück—An Appreciation," *The Amer. Scholar*, 51(3):351–367.

Hadamard, J. (1949). "The Psychology of Invention in the Mathematical Field," Princeton, N.J.: Princeton University Press.

Hahn, O. and Strassmann, F. (1939). "Über den Nachweis und das Verhalten bei der Bestrahlung des Urans mittels Neutronen entstehenden Erdalkalimetalle," *Naturwissensch.*, 27:11–15. "Nachweis der Entstehung aktiver Bariumisotope aus Uran und Thorium durch Neutronenbestrahlung; Nachweis weiterer aktiver Bruchstücke bei der Uranspaltung," *Naturwissensch.*, 27:89–95.

Hahn, O. (1968). *Mein Leben*. Munich: *Naturwissensch.*, 27:89–95.

Harm, H. (1976). "Repair of UV-Irradiated Biological Systems: Photoreactivation," in S. Y. Wang, ed., *Photochemistry and Photobiology of Nucleic Acids*. Vol. 2, New York: Academic Press, pp. 219–263.

Harnack, A. von, (1950). *Das Wesen des Christentums* (Berlin, 1900). 15th edition, Leipzig: Himmihs Verlag.

Hassenstein, B., and Reichhardt, W. (1953). "Der Schluss von Reiz-Reaktions-Funktionen auf Systemstrukturen," *Zeitschr. Naturforsch.*, 8b:518–524.

Hayashi, I., Larner, J., and Sato, G. (1978). "Hormonal Growth Control of Cells in Culture," *In vitro*, 14:23–30.

Heisenberg, W. (1925). "Über quantentheoretische Umdeutung kinematischer und mechanischer Beziehungen," *Zeitschr. f. Physik*, 33:879.

———— (1927). "Über den anschaulichen Inhalt der quantentheoretischen Kinematik und Mechanik," *Zeitschr. f. Physik*, 43:172–198.

———— (1969). *Der Teil und das Ganze*. Munich: Piper Verlag.

Heitler, W., and London, F. (1927). "Wechselwirkung neutraler Atome und homöopolare Bindung nach der Quantenmechanik," *Zeitschr. f. Physik*, 44:455–472.

Herelle, F. d' (1926). *The Bacteriophage and Its Behavior*. Baltimore: Williams and Wilkins.

Herriott, R. M. (1983). "John H. Northrop: The Nature of Enzymes and Bacteriophage," *Trends in Bioch. Sci.*, 8 (August):296–297.

Hershey, A. (1946). "Mutation of Bacteriophage with Respect to Type of Plaque," *Genetics*, 31:620–640.

————, and Rotman, R. (1949). "Genetic Recombination Between Host-range and Plaque-type Mutants of Bacteriophage in Single Bacterial Cells," *Genetics*, 34:44–71.

————, and Chase, M. (1952). "Independent Functions of Viral Protein and Nucleic Acid in Growth of Bacteriophage," *J. Gen. Physiol.*, 36: 39–56.

———— (1966). "The Injection of DNA into Cells by Phage," in PATOOMB, 100–108.

———— (1970). "Genes and Hereditary Characteristics," *Nature*, 226: 697–700.

———— (1981). In the Max Delbrück Dedication Ceremony, Cold Spring Harbor Laboratory, Cold Spring Harbor, N.Y.

Hildebrand, E., and Dencher, N. (1975). "Two Photosystems Controlling Behavioural Responses in H. halobium," *Nature*, 257:46–48.

Hinshelwood, C. N. (1950). "Bacteriology and Chemical Kinetics," *Endeavour*, 8:151.

Hofstadter, D. R. (1979). *Gödel, Escher, Bach*. New York: Basic Books.

Holton, G. (1973). "The Roots of Complementarity," in *Thematic Origins of Scientific Thought: Kepler to Einstein*. Cambridge, Mass.: Harvard University Press.

Horowitz, N. H. (1950). "Biochemical Genetics of Neurospora," *Adv. in Genetics*, 3:33–71.

—— (1973). "Neurospora and the Beginnings of Molecular Genetics," *Neurosp. Newsletter*, 20 (June).

—— (1979). "Genetics and the Synthesis of Proteins," *Ann. New York Acad. Sci.*, 325:253–266.

——, and Leupold, U. (1951). "Some Recent Studies Bearing on the One Gene–One Enzyme Hypothesis," *Cold Spring Harbor Symp. on Quantitative Biol.*, Vol. XVI ("Genes and Mutations"), 65–75.

Jackson, S., and Hinshelwood, C. N. (1950). "An Investigation of the Nature of Certain Adaptive Changes in Bacteria," *Proc. Royal Soc.*, B 136:562–576.

Jauch, J. (1973). *Are Quanta Real?* Bloomington, Ind.: Indiana University Press.

Jerne, N. K. (1966). "The Natural Selection Theory of Antibody Formation: Ten Years Later," in PATOOMB, 301–312.

Johannsen, I. (1980). *Meilensteine der Genetik*. Hamburg: Parey-Verlag.

Johannsen, W. L. (1909). *Elemente der exakten Erblichkeitslehre*. Jena: Fischer-Verlag.

Jordan, P. (1932). "Die Quantenmechanik und die Grundprobleme der Biologie und Psychologie," *Naturwissensch.*, 20:815–821.

—— (1934). "Quantenphysikalische Bemerkungen zur Biologie und Psychologie," *Erkenntnis*, 4:215–252.

—— (1938). "Zur Frage einer spezifischen Anziehung zwischen Genmolekülen," *Phys. Zeitschr.*, 39:711–714.

—— (1939). "Über quantenmechanische Resonanzanziehung und über das Problem der Immunitätsreaktion," *Zeitschr. f. Physik*, 113:431–438.

—— (1941). *Die Physik und das Geheimnis des organischen Lebens*. Braunschweig: Vieweg-Verlag.

—— (1944). "Quantenphysik und Biologie," *Naturwissensch.*, 32:309–316.

Judson, H. F. (1979). *The Eighth Day of Creation*. New York: Simon and Schuster.

Kaiser, D., and Manoil, C. (1979). "Myxobacteria: Cell Interactions, Genetics, and Development," *Ann. Rev. Microbiol.*, 33:595–639.

Kaissling, K. E., and Priester, E. (1970). "Die Riechschwelle des Seidenspinners," *Naturwissensch.*, 57:23–28.

Kalmanson, G. M., and Bronfenbrenner, J. J. (1939). "Studies on the Purification of Bacteriophage," *J. Gen. Physiol.*, 23:203–228.

Kay, L. E. (1985). "Conceptual Models and Analytical Tools: The Biology of the Physicist Max Delbrück 1931–1946," *J. Hist. Biol.*, 812:207–246.

Kelner, A. (1949). "Effects of Visible Light on the Recovery of *Streptomyces griseus* conidia from Ultraviolet Irradiation Injury," *Proc. Natl. Acad. Sci. U.S.A.*, 35:73–79.

Kendrew, J. C. (1967). "How Molecular Biology Started," *Scientific American,* 216 (March):141–143.

Kluyver, A. J., and Donker, H. J. L. (1926). "The Unity of Biochemistry," *Chem. Zelle Gewebe,* 13:134–190.

———, and Van Niel, C. B. (1956). *The Microbe's Contribution to Biology.* Cambridge, Mass.: Harvard University Press.

Kohler, R. E. (1976). "The Management of Science: The Experience of Warren Weaver and the Rockefeller Foundation Programme in Molecular Biology," *Minerva,* 14:279–306.

Kornberg, A. (1974). *DNA Synthesis.* San Francisco: W. H. Freeman.

Krafft, F. (1982). *Das Selbstverständnis der Physik im Wandel der Zeit.* Weinheim: Physik-Verlag.

Kuhn, T. (1962). *The Structure of Scientific Revolutions.* Chicago: University of Chicago Press.

——— (1978). *Black-Body Theory and the Quantum Discontinuity.* Oxford: Clarendon Press.

Lederberg, J., and Tatum, E. (1946a). "Novel Genotypes in Mixed Cultures of Biochemical Mutants in Bacteria," *Cold Spring Harbor Symp. on Quantitative Biol.,* Vol. XI ("Heredity and Variation in Microorganisms"), 113–114.

——— (1946b). "Genetic Recombination in E. coli," *Nature,* 158:558.

Leppmann, W. (1981). *Rilke.* Bern: Scherz Verlag.

Lewis, I. M. (1934). "Bacterial Variation with Special Reference to Some Mutable Strains of Colon Bacteria in Synthetic Media," *J. Bacteriol.,* 28:- 619–640.

Lipson, E. D. (1975). "White Noise Analysis of *Phycomyces,*" *Biophys. J.,* 15:1013–1031.

———, and Block, S. M. (1983). "Light and Dark Adaptation in *Phycomyces,*" *J. Gen. Physiol.,* 81:845–859.

Locke, M. (1965). "Permeability of Insect Cuticle to Water and Lipids," *Science,* 147:295–298.

Lukacs, G. (1954). *Die Zerstörung der Vernuft.* Berlin, (1973) new edn. Neuwied: Luchterhand Verlag.

Luria, S. E. (1945). "Mutations of Bacterial Viruses Affecting Their Host Range," *Genetics,* 30:84–91.

——— (1947). "Reactivation of Irradiated Bacteriophage by Transfer of Self-reproducing Units," *Proc. Natl. Acad. Sci. U.S.A.,* 33:253–264.

——— (1950). "Bacteriophage: An Essay on Virus Reproduction," *Science,* 111:507–509.

——— (1966). "Mutations of Bacteria and Bacteriophage," in PATOOMB.

——— (1984). *A Slot Machine, A Broken Test Tube.* New York: Harper & Row.

———, and Anderson, T. F. (1942). "The Identification and Characterization of Bacteriophages with the Electron Microscope," *Proc. Natl. Acad. Sci. U.S.A.,* 28:127–130.

———, and Latarjet, R. (1947). "Ultraviolet Irradiation of Bacteriophage During Intracellular Growth," *J. Bacteriol.,* 53:149–163.

Lwoff, A. (1953). "Lysogeny," *Bacteriol. Rev.,* 17:269–337.

——— (1966). "The Prophage and I," in PATOOMB, 88–99.

Maddox, J. (1983). "Good Cause for Celebration," *Nature,* 305:177.

McClintock, B. (1930). "A Cytological Demonstration of an Interchange

Between Two Non-homologous Chromosomes of *Zea mays,*" *Proc. Natl. Acad. Sci. USA,* 16:791–796.

—— (1934). "The Relation of a Particular Chromosomal Element to the Development of the Nucleoli in *Zea. mays,*" *A. Zellforsch. u. Mikr. Anat.,* 21:294–328.

—— (1961). "Some Parallels Between Gene Control Systems in Maize and in Bacteria," *American Naturalist,* 95:266.

McCormack, R. (1982). *Night Thoughts of a Classical Physicist.* Cambridge, Mass.: Harvard University Press.

Medvedev, Z. A. (1971). *Der Fall Lysenko.* Hamburg: Hoffmann und Campe.

Meitner, L. (1925). "Die Gamma-Strahlung der Actiniumreihe und der Nachweis, dass die Gamma-Strahlen erst nach erfolgtem Atomzerfall emitiert werden," *Zeitschr. f. Physik,* 34:807.

——, and Phillip, K. (1931). "Das Gamma-Spektrum von Th C und die Gamowsche Theorie der Alpha-Feinstruktur," *Naturwissensch.,* 19:1009.

——, and Hahn, O. (1936). "Neue Umwandlungsprozesse bei Bestrahlung des Urans mit Neutronen," *Naturwissensch.,* 24:158.

Melchers, G. (1965). "Biologie und Nationalsozialismus," in A. Flitner, ed., *Deutsches Geistesleben und Nationalsozialismus.* Tübingen: Wunderlich Verlag.

Meselson, M., and Stahl, F. (1958). "The Replication of DNA in E. coli," *Proc. Natl. Acad. Sci. U.S.A.,* 44:413.

Meyenn, K. von (1982). "Pauli, das Neutrino und die Entdeckung des Neutrons vor 50 Jahren," *Naturwissensch.,* 69:564–573.

Mitchell, H. K., and Houlahan, M. B. (1946). "Adenine-requiring Mutants of Neurospora crassa," *Fed. Proc.,* 5:370–375.

Mitchell, P. (1961). "A Chemiosmotic Hypothesis for the Mechanism of Oxidative Phosphorylation," *Nature,* 191:144.

Moore, Ruth. (1974). *Niels Bohr.* Cambridge, Mass.: MIT Press. (Paperback 1985.)

Morgan, T. H. (1940). "Calvin B. Bridges (1889–1938)," *Biogr. Mem. of the Natl. Acad. Sci. U.S.A.,* Vol. XXII, 31–48.

Mueller, P., and Rudin, D. O. (1967). "Action Potential Phenomena in Experimental Bimolecular Lipid Membranes," *Nature,* 213:603–604.

—— (1968). "Action Potentials Induced in Bimolecular Lipid Membranes," *Nature,* 217:713–719.

Muller, H. J. (1927). "Artificial Transmutations of the Gene," *Science,* 66:84–87.

—— (1928). "The Production of Mutations by X-rays," *Proc. Natl. Acad. Sci. U.S.A.,* 14:714–726.

—— (1936). "The Need of Physics in Attack on Fundamental Problems of Genetics," *Scientific Monthly,* 44:210–214.

—— (1940). "An Analysis of the Process of Structural Change in Chromosomes of Drosophila," *J. of Genetics,* 40:1–66.

Mullins, N. C. (1972). "The Development of a Scientific Speciality: The Phage Group and the Origins of Molecular Biology," *Minerva,* 10:51–82.

Neher, E., and Sakmann, B. (1976). "Single-channel Currents Recorded from Membrane of Denervated Frog Muscle Fiber," *Nature,* 260:799–801.

Neumann, J. von (1966). *Theory of Self-Reproducing Automata,* edited and completed by A. W. Burk. Urbana, Ill.: University of Illinois Press.

Northrop, J. H. (1939). *Crystalline Enzymes, The Chemistry of Pepsin, Trypsin and Bacteriophage.* New York: Columbia University Press.

Novikoff, A. B. (1945). "The Concept of Integrative Levels in Biology," *Science,* 101:209–215.

Oesterhelt, D., and Stoeckenius, W. (1971). "Rhodopsin-like Protein from the Purple Membrane of H. halobium," *Nature New Biology,* 233:149–152.

Olby, R. C. (1974). *The Path to the Double Helix.* London: Macmillan.

Oort, A. J. P. (1931). "The Spiral Growth of *Phycomyces,*" *Proc. Roy. Acad. Sci. Amsterdam,* 34:564–575.

Ootaki, T., Fischer, E. P., and Lockhardt, P. (1974). "Complementation Between Mutants of *Phycomyces* with Abnormal Phototropism," *Mol. Gen. Genet.,* 131:233–246.

Pauk, W. (1968). *Harnack and Troeltsch.* New York: Oxford University Press.

Pauling, L., and Corey, R. B. (1952). "Configuration of Polypeptide Chains with Equivalent Cis Amide Groups," *Proc. Natl. Acad. Sci. U.S.A.,* 38:86–89.

——— (1953). "A Proposed Structure for the Nucleic Acids," *Proc. Natl. Acad. Sci. U.S.A.,* 39:84–97.

Planck, M. (1900). "Über eine Verbesserung der Wienschen Spektralgleichung," *Verh. Dtsch.phys.Ges. Berlin,* 2:202.

Plessner, H. (1927). *Die Stufen des Organischen und der Mensch.* Stuttgart: Göschen.

Poisson, S. D. (1837). Paris: *Recherches sur la probabilité des jugements en matière criminelle et en matière civile, precédées des règles générales du calcul des probabilités.*

Polya, G. (1921). "Über eine Aufgabe der Wahrscheinlichkeitsrechnung betreffend die Irrfahrt im StraBennetz," *Math. Ann.,* 84:149–160.

Pontecorvo, G. (1958). *Trends in Genetic Analysis.* New York: Columbia University Press.

Poo, M., and Cone, R. A. (1974). "Lateral Diffusion of Rhodopsin in the Photoreceptor Membrane," *Nature,* 247:438–441.

Popper, K. R. (1966). *The Logic of Scientific Discovery.* London: Hutchinson.

——— (1983). *Realism and the Aim of Science.* London: Hutchinson/ Rowman & Littlefield.

Primas, H. (1981). *Chemistry, Quantum Mechanics, and Reductionism.* Berlin: Springer Verlag.

Quinn, W. G., and Gould, J. L. (1979). "Nerves and Genes," *Nature,* 278:19–23.

Racker, E. (1982). "Otto Warburg at a Turning Point in 1932," *Trends in Bioch. Sci.* (December), 448–449.

Reichardt, W. (1966). "Cybernetics of the Insect Optomotor Response," in PATOOMB, 313–334.

———. (1978). "Functional Characterization of Neural Interaction Through an Analysis of Behavior," in F. O. Schmitt, ed., *The Neurosciences: Fourth Study Program.* New York: Rockefeller University Press, pp. 81–103.

Reif, F. (1965). *Fundamentals of Statistical and Thermal Physics.* New York: McGraw-Hill.

Rohrlich, F., and Gluckstern, R. L. (1952). "Forward Scattering of Light by a Coulomb Field," *Physical Rev.*, 86:1–9.

Rosenfeld, L. (1963). "Niels Bohr's Contribution to Epistemology," *Physics Today*, 16:47–52.

Ruska, H. (1940). "Die Sichtbarmachung der Bakteriophagen Lyse im Übermikroskop," *Naturwissensch.*, 28:45–46.

—— (1941). "Über ein neues bei der Bakteriophagenlyse auftretendes Formelement," *Naturwissensch.*, 29:367–368.

Schilpp, P. A., ed. (1949). "Albert Einstein, Philosopher-Scientist," *The Library of Living Philosophers*, Vol. VII. La Salle, Ill.: Open Court.

Schrödinger, E. (1926). "Quantisierung als Eigenwertproblem," *Ann. Phys.*, 79:361.

—— (1945). *What Is Life?* Cambridge, Engl.: Cambridge University Press; reprint, 1967, including "Mind and Matter."

Setlow, R. B., and Pollard, E. C. (1962). "Action Spectra and Quantum Yields," in Setlow, ed., *Molecular Biophysics*. Oxford: Pergamon Press, pp. 276–305.

Shannon, C. (1948). "A Mathematical Theory of Communication," *The Bell System Techn. Journ.*, 27:379–423 and 623–656.

Singer, S. J., and Nicholson, G. L. (1972) "The Fluid Mosaic Model of the Structure of Cell Membranes," *Science*, 175:720–721.

Singer, S. J. (1974). "The Molecular Organization of Membranes," *Ann. Rev. Biochem.*, 43:805–834.

Speiser, A. (1937). *Die Theorie der Gruppen von endlicher Ordnung.* 3rd edn., Berlin: Julius Springer Verlag.

Sonneborn, T. (1949). "Beyond the Gene," *Amer. Scientist*, 37: 33–59.

Spengler, O. (1918). *The Decline of the West* (English trans. 1926) . New York.

Stadler, L. (1939). "Genetic Studies with Ultraviolet Light," *Proc. of VII Int. Conf. Genet.*, 269–276.

Stanley, W. M. (1935). "Isolation of a Crystalline Protein Possessing the Properties of Tobacco Mosaic Virus," *Science*, 81:644–645.

Stent, G. S. (1963). *Molecular Biology of Bacterial Viruses.* San Francisco: W. H. Freeman.

—— (1968). "That Was the Molecular Biology That Was," *Science*, 160:390–395.

—— (1966). "Waiting for the Paradox," in PATOOMB, 3–8.

—— (1969). *The Coming of the Golden Age.* New York: Doubleday, Natural History Press.

—— (1978). *Paradoxes of Progress.* San Francisco: W. H. Freeman.

—— (1979). "Naturwissenschaft und Ethik als paradoxe Schöpfungen der Vernunft," *Naturwissensch.*, 66:354–357.

——, and Calendar, R. (1978). *Molecular Genetics: An Introductory Narrative.* San Francisco: W. H. Freeman.

——. (1981). "Obituary: Max Delbrück," *Trends in Biochem. Sciences*, 6:iii–iv.

Sturtevant, A. H. (1913). "The Linear Arrangement of Six Sex-linked Factors in Drosophila, as Shown by Their Mode of Association," *J. Exp. Zool.*, 14:43–59.

——, and Beadle, G. W. (1939). *An Introduction to Genetics.* Philadelphia: W. B. Saunders.

Sturtevant, A. H. (1965). *A History of Genetics.* New York: Harper & Row.

Sutton, W. S. (1903). "The Chromosomes in Heredity," *Biol. Bull.,* 4:213–215.

Tarski, A. (1969). "Truth and Proof," *Scientific American,* 220 (June), 63–77.

Thimme, A. (1955) *Hans Delbrück als Kritiker der Wilhelminischen Epoche.* Düsseldorf: Droste Verlag.

Timoféeff-Ressovsky, N. W., and Zimmer, K. G. (1947). *Das Trefferprinzip in der Biologie.* Stuttgart: Hirzel Verlag.

Watson, J. D. (1966). "Growing Up in the Phage Group," in PATOOMB, 239–245.

——. (1968). *The Double Helix.* New York: Atheneum.

—— (1975). *Molecular Biology of the Gene.* 3rd edn., Menlo Park, N.J.: Benjamin.

Weber, E. H. (1846). "Tastsinn und Gemeingefühl," *Nachdruck als Ostwalds Klassiker der exakten Wissenschaften,* No. 149, Leipzig (1905).

Weidel, W. (1956). *Virus und Molekular-Biologie.* Berlin: Springer Verlag.

Weizsäcker, C. F. von (1963). *Zum Weltbild der Physik.* Stuttgart: Hirzel Verlag.

—— (1983). *Die Wahrnehmung der Neuzeit.* Munich: Hanser Verlag.

Wiener, N. (1947). *Cybernetics.* Cambridge, Mass.: MIT Press.

Wigner, E. (1927). "Einige Folgerungen aus der Schrödingerschen Theorie für die Termstrukturen," *Zeitschr. f. Physik,* 43:624–652.

Yoxen, E. J. (1979). "Where Does Schrödinger's *What Is Life?* Belong in the History of Molecular Biology?", *History of Science,* 17:17–52.

Yukawa, H. (1982). *Tabibito.* Cleveland: World Scientific Publishing Company.

Zimmer, K. G. (1966). "The Target Theory," in PATOOMB, 33–42.

Bibliography of Works by Max Delbruck (1906–1981)

1   1928   "Ergänzung zur Gruppentheorie der Terme," *Z. Phys.,* 51:181–187.

2   1930   "Quantitatives zur Theorie der homöopolaren Bindung," *Ann. Phys.,* 5:36–58.

3   1930   "The Interaction of Inert Gases," *Proc. Royal Soc. London.,* A 129:686–698.

4   1931   with George Gamow
"Übergangswahrscheinlichkeiten von angeregten Kernen," *Z. Phys.,* 72:-492–499.

5   1932   "Possible Existence of Multiply Charged Particles of Mass One," *Nature,* 130:626–627 (on p. 660).

6   1933   "Zusatz bei der Korrektur," *Z. Phys.,* 84: 144, in L. Meitner and H. Köster, "Über Streuung kurzwelliger Gamma-Strahlen," *Z. Phys.,* 84:-137–144.

7   1935   with Lise Meitner
Der Aufbau der Atomkerne. Natürliche und künstliche Kernumwandlungen. Berlin: Julius Springer Verlag.

8   1935   with N. W. Timoféeff-Ressovsky and K. G. Zimmer
"Über die Natur der Genmutation und der Genstruktur," *Nachr. Ges. Wiss. Göttingen, Math.-Phys. Kl.*, Fachgruppe 6, No. 13, 190–245.

9   1936   with N. W. Timoféeff-Ressovsky
"Strahlengenetische Versuche über sichtbare Mutationen und die Mutabilität einzelner Gene bei *Drosophila melanogaster*," *Z. Indukt. Abstamm. Vererbungsl.* 71:322–334.

10   1936   with N. W. Timoféeff-Ressovsky
"Cosmic Rays and the Origin of Species," *Nature*, 137:358–359.

11   1936   with Gerd Molière
"Statistische Quantenmechanik und Thermodynamik," *Abh. Preuss. Akad. Wiss. Phys. I Math. Kl.*, 1:1–46.

12   1939   with Emory L. Ellis
"The Growth of Bacteriophage," *J. Gen. Physiol.*, 22:365–384.

13   1940   "Statistical Fluctuations in Autocatalytic Reactions," *J. Chem. Phys.*, 8:120–124.

14   1940   "Radiation and the Hereditary Mechanism," *Amer. Naturalist*, 74:350–362.

15   1940   "The Growth of Bacteriophage and Lysis of the Host," *J. Gen. Physiol.*, 23:643–660.

16   1940   "Adsorption of Bacteriophage Under Various Physiological Conditions of the Host," *J. Gen. Physiol.*, 23:631–642.

17   1940   with Linus Pauling
"The Nature of Intermolecular Forces Operative in Biological Processes," *Science*, 92:77–79.

18   1940   "Growth of Bacteriophage and Lysis of the Host," *J. Tenn. Acad. Sci.*, 15:417 (abstract).

19   1941   "A Theory of Autocatalytic Synthesis of Polypeptides and Its Application to the Problem of Chromosome Reproduction," *Cold Spring Harbor Symp. on Quantitative Biol.*, Vol. IX ("Genes and Chromosomes"), 122–124.

20   1942   with S. E. Luria
"Interference Between Bacterial Viruses. I. Interference Between Two Bacterial Viruses Acting upon the Same Host, and the Mechanism of Virus Growth," *Arch. Bioch.*, 1:111–141.

21   1942   with S. E. Luria
"Interference Between Inactivated Bacterial Virus and Active Virus of the Same Strain and of a Different Strain," *Arch. Bioch.*, 1:207–218.

22   1942   "Bacterial Viruses (Bacteriophages)," *Adv. Enzymol.*, 2:1–32.

23   1942   "The Reproduction of Bacteriophage," *J. Bacteriol.*, 45:74 (abstract).

24   1943   with S. E. Luria and T. F. Anderson
"Electron Microscope Studies of Bacterial Viruses," *J. Bacteriol.*, 46:57–77.

25   1943   with S. E. Luria
"Mutations of Bacteria from Virus Sensitivity to Virus Resistance," *Genetics*, 28:491–511.

26   1943   with S. E. Luria
"A Comparison of the Action of Sulfa-Drugs on the Growth of a Bacterial Virus and of Its Host," *J. Bacteriol.*, 46:574–575, and
1944 *Proc. Indiana Acad. Sci.*, 53:28–29 (abstract).

27   1944   "A Statistical Problem," *J. Tenn. Acad. Sci.*, 19:177–178.

28 1945 "Spontaneous Mutations of Bacteria," *Ann. Mo. Bot. Gard.,* 32:223–233.

29 1945 "The Burst Size Distribution in the Growth of Bacterial Viruses (Bacteriophages)," *J. Bacteriol.,* 50:131–135.

30 1945 "Effects of Specific Antisera on the Growth of Bacterial Viruses," *J. Bacteriol.,* 50:137–150.

31 1945 "Interference Between Bacterial Viruses. III. The Mutual Exclusion Effect and the Depressor Effect," *J. Bacteriol.,* 50:151–170.

32 1945 with T. F. Anderson and M. Demerec
"Types of Morphology Found in Bacterial Viruses," *J. Appl. Phys.,* 16:264.

33 1946 "Experiments with Bacterial Viruses (Bacteriophages)," Harvey Lectures 41:161–187.

34 1946 "Bacterial Viruses or Bacteriophages," *Biol. Rev.,* 21:30–40.

35 1946 with W. T. Bailey, Jr.
"Induced Mutations in Bacterial Viruses," *Cold Spring Harbor Symp. Quantitative Biol.,* Vol. XI "Heredity (and Variation in Microorganisms"), pp. 33–37.

36 1947 "Über Bakteriophagen," *Naturwissensch.,* 34:301–306.

37 1948 "Biochemical Mutants of Bacterial Viruses," *J. Bacteriol.,* 56:1–16.

38 1948 with Mary Bruce Delbrück
"Bacterial Viruses and Sex," *Scientific American,* 179 (November):46–51; German translation in
1949 *Naturwiss. Rundschau,* 7:301–306.

39 1949 "Génétique du Bacteriophage," *Colloq. Int. C.N.R.S.,* 8:91–103.

40 1949 *Symposium sur Unités Biologiques Donées de Continuité Génétique.* Paris: Editions du Centre National de la Recherche Scientifique, pp. 33–35.

41 1949 "A Physicist Looks at Biology," *Trans. Conn. Acad. Arts Sci.,* 38:173–190; reprinted in J. Cairns, G. S. Stent, and J. D. Watson, eds., *Phage and the Origins of Molecular Biology.* Cold Spring Harbor, N.Y. (1966) (translations in German, Danish, and Japanese).

42 1950 editor of *Viruses 1950.* Proceedings of Conference held at the California Institute of Technology, Pasadena, March 20–22, 1950.

43 1951 with Roderick K. Clayton
"Purple Bacteria," *Scientific American,* 185 (November):68–72.

44 1952 with J. J. Weigle
"Mutual Exclusion Between an Infecting Phage and a Carried Phage," *J. Bacteriol.,* 62:301–318.

45 1953 with N. Visconti
"The Mechanism of Genetic Recombination in Phage," *Genetics,* 38:5–33.

46 1954 "Wie vermehrt sich ein Bakteriophage?", *Angew. Chemie,* 66:391–395.
in English: 1956 "Current Views on the Reproduction of Bacteriophage," *Scientia,* 91:118–126.

47 1954 "On the Replication of Desoxyribonucleic Acid (DNA)," *Proc. Natl. Acad. Sci. U.S.A.,* 40:783–788.

48 1956 with W. Reichhardt
"System Analysis for the Light Growth Reactions of *Phycomyces,"* in Doro-

thea Rudnick, ed., *Cellular Mechanisms in Differentiation and Growth.*
Princeton, N.J.: Princeton University Press, pp. 3–44.
    49   1957   with G. S. Stent
"On the Mechanism of DNA Replication," in W. D. McElroy and B. Glass,
eds., *The Chemical Basis of Heredity.* Baltimore: Johns Hopkins Press.
    50   1958   *Bacteriophage Genetics,* Proc. of the Fourth Intern. Polio-
myelitis Congress, New York: Lippincott.
    51   1958   with S. W. Golomb and L. R. Welch
"Construction and Properties of Comma-Free Codes," *Biol. Medd. Dan.
Vid. Selsk.,* 23(9):1–30.
    52   1958   with R. Cohen
"Distribution of Stretch and Twist Along the Growing Zone of the Sporan-
giophore of *Phycomyces* and the Distribution of Response to a Periodic
Illumination Program," *J. Cell. Comp. Physiol.,* 52:361–388.
    53   1959   with R. Cohen
"Photoreactions in *Phycomyces:* Growth and Tropic Responses to the Stim-
ulation of Narrow Test Areas," *J. Gen. Physiol.,* 42:677–695.
    54   1960   with W. Shropshire, Jr.
"Action and Transmission Spectra of *Phycomyces,*" *Plant Physiol.,* 35:194–
204.
    55   1961   with D. Varju
"Photoreactions in *Phycomyces:* Responses to the Stimulation of Narrow
Test Areas with UV Light," *J. Gen. Physiol.,* 44:1177–1188.
    56   1961   with D. Varju and L. Edgar
"Interplay Between the Reactions to Light and to Gravity in *Phycomyces,*"
J. Gen. Physiol., 45:47–58.
    57   1962   "Knotting Problems in Biology," *Proc. Symp. Appl. Math.,*
14:55–68.
    58   1962   with H. E. Johns and S. A. Rapaport
"Photochemistry of Thymine Dimers," *J. Mol. Biol.,* 4:104–114.
    59   1962   "Genetik und die Synthese 'lebender' Substanz," *Der Mathem.
und Naturwiss. Unterr.,* 15:241–243.
    60   1962   "Ein Hinweis auf einige neue Gedanken in der Biologie,"
Physikalische Blätter, 18:559–562.
    61   1963   "Der Lichtsinn von *Phycomyces,*" *Ber. Dtsch. Bot. Ges.,* 75:-
411–430.
    62   1963   "Biophysics," Commemoration of the 50th Anniversary of
Niels Bohr's Papers on Atomic Constitution, Session on Cosmos and Life,
Institute for Theoretical Physics, Copenhagen, pp. 41–67.
    63   1963   "Inwiefern ist die Biologie zu schwierig für die Biologen?",
Physikertagung Stuttgart. Moosbach/Baden; Physik-Verlag.
    64   1963   "Über Vererbungschemie," *Arbeitsgemeinschaft Forsch.
Land. Nordrhein-Westfalen, Nat.-Ing.-Gesellschaft, Wiss. Veröff.,* No. 125:
1–39.
    65   1963   "Die Vererbungschemie," *Naturwiss. Rundschau,* 16:85–
89.
    66   1963   "Das Begriffsschema der Molekular-Genetik," *Nova Acta Leo-
poldina,* N.F.26, No. 165:9–16. In parts also as
1964 "Betrachtungen über die Soziologie der Vererbungschemie," *Das
Leben,* 1(4):136–138.
    67   1965   "Primary Transduction Mechanisms in Sensory Physiology

and the Search for Suitable Experimental Systems," *Israel J. Med. Sci.,* 1:1363–1365.

68 1966 with others
"General Discussion," *Radiat. Res. Suppl.,* 6:227–234.

69 1967 "Molecular Aspects of Genetics," in R. A. Brink, ed., *Heritage from Mendel.* Madison, Wis.: University of Wisconsin Press.

70 1967 with K. L. Zankel and P. V. Burke
"Absorption and Screening in *Phycomyces,"J. Gen. Physiol.,* 50:1893–1906.

71 1968 with G. Adam
"Reduction of Dimensionality in Biological Diffusion Processes," in N. Davidson and A. Rich, eds., *Structural Chemistry and Molecular Biology* (Linus Pauling Festschrift). San Francisco: W. H. Freeman, pp. 198–215.

72 1968 "Molecular Biology—The Next Phase," *Eng. Sci.,* 32 (November):36–40.

in French:
"Biologie moléculaire: La prochaine étape," *Sciences* (Paris) 56:7–13.

73 1968 with G. Meissner
"Carotenes and Retinal in *Phycomyces,"* *Plant Physiol.,* 43:1279–1283.

74 1969 with K. Bergman, Patricia V. Burke, E. Cerdá-Olmedo, C. N. David, K. W. Foster, E. W. Goodell, M. Heisenberg, G. Meissner, M. Zalokar, D. S. Dennison, and W. Shropshire, Jr., *"Phycomyces,"Bacteriol. Rev.,* 33:99–157.

75 1969 with R. Edgar
"Jean Jacques Weigle 1901–1968," *Eng. Sci.,* 33 (January):21.

76 1970 "A Physicist's Renewed Look at Biology: Twenty Years Later," *Science,* 168:1312–1315; also in *Les Prix Nobel en 1969.* Stockholm: The Nobel Foundation.

77 1970 "Lipid Bilayers as Models of Biological Membranes," in F. O. Schmitt, ed., *The Neurosciences: Second Study Program.* New York: Rockefeller University Press, pp. 677–684.

78 1970 with M. Petzuch
"Effects of Cold Periods on the Stimulus-Response System of *Phycomyces,"* *J. Gen. Physiol.,* 56:297–308.

79 1970 "Vorwort," in E. Geissler, ed., *Desoxyribonukleinsäure.* Berlin: Akademie Verlag.

80 1971 "Aristotle-totle-totle," in J. Monod and E. Borek, eds., *Of Microbes and Life* (Andre Lwoff Festschrift). New York: Columbia University Press.

81 1972 "Homo Scientificus According to Beckett," in W. Beranek, ed., *Science, Scientists and Society.* New York: Bodgen and Quigley; German translation *Neue Sammlung,* 12:528–542.

82 1972 "Signal Transducers: *Terra Incognita* of Molecular Biology," *Angew. Chemie International Edition in English,* 11:1–7.

83 1972 "Geleitwort zur deutschen Ausgabe," in E. Geissler, ed., *Phagen und die Entwicklung der Molekularbiologie.* Berlin: Akademie Verlag.

84 1972 "Out of This World," in F. Reines, ed., *Cosmology, Fusion and Other Matters* (George Gamow Memorial Volume). Boulder, Col.: Associated University Press.

85 1972 with Carol Lipson and Edward Lipson
"Anfänge der Wahrnehmung," *Mannheimer Forum,* 72:53–84. H. v. Ditfurth, ed. (publ. by Boehringer-Mannheim).

86 1973 with T. Ootaki, A. Crafts Lighty, and W. J. Hsu
"Complementation Between Mutants of *Phycomyces* Deficient with Respect to Carotenogenesis," *Mol. Gen. Genet.*, 121:57–70.

87 1974 "Anfänge der Wahrnehmung: Untersuchungen über den Mechanismus der Wandlung von Sinnessignalen bei *Phycomyces,*" Karl-August-Forster-Lectures, 10 (1973). Mainz: Akademie der Wissenschaften und der Literatur.

88 1974 with Wan-Jean Hsu and David C. Ailion
"Carotenogenesis in *Phycomyces,*" *Phytochemistry*, 13:1463–1468.

89 1974 Interview
"Education for Suicide," *Prism* (November): 16–19ff.

90 1975 with R. J. Cohen, Y. N. Jan, and J. Matricon
"Avoidance Response, House Response, and Wind Response of the Sporangiophore of *Phycomyces,*" *J. Gen. Physiol.*, 66:67–95.

91 1975 with A. P. Eslava, M. I. Alvarez, and P. V. Burke
"Genetic Recombination in Sexual Crosses of *Phycomyces,*" *Genetics*, 80:-445–462.

92 1975 with P. G. Saffman
"Brownian Motion in Biological Membranes," *Proc. Natl. Acad. Sci. U.S.A.*, 72:3111–3113.

93 1975 with A. P. Eslava and M. I. Alvarez
"Meiosis in *Phycomyces,*" *Proc. Natl. Acad. Sci. U.S.A.*, 72:4076–4080.

94 1976 with A. Katzir and D. Presti
"Responses of *Phycomyces* Indicating Optical Excitation of the Lowest Triplet State of Riboflavin," *Proc. Natl. Acad. Sci. U.S.A.*, 73:1969–1973.

95 1976 "Light and Life III," *Carlsberg Res. Commun.*, 41: 299–309.

96 1976 "How Aristotle Discovered DNA," in K. Huang, ed., *Physics and Our World: A Symposium in Honor of Victor F. Weisskopf.* New York: American Institute of Physics.

97 1977 with D. Presti and W. J. Hsu
"Phototropism in *Phycomyces* Mutants Lacking Beta-Carotene," *Photochem. Photobiol.*, 26:403–405.

98 1977 with E. Cerdá-Olmedo
"El Comportamiento de *Phycomyces,*" *Genética Microbiana.* Madrid: Editorial Alhambra.

99 1978 "Erinnerungen an Max Born," in H. Baumann, ed., *Max-Born-Gymnasium Germering: Jahresbericht 1977–1978.*

100 1978 with David Presti
"Photoreceptors for Biosynthesis, Energy, Storage and Vision," *Plant, Cell, and Environment*, 1:81–100.

101 1978 "Mind from Matter?", *American Scholar*, 47:339–353.

102 1978 "Mind from Matter?", in W. H. Heidcamp, ed., *The Nature of Life.* Baltimore: University Park Press.

103 1978 "The Arrow of Time—Beginning and End," *Eng. Sci.*, 42 (September):5–9.

104 1978 "Virology Revisited," *Proc. of International Symposium on Molecular Basis of Host Virus Interaction.* Princeton, N.J.: Science Press.

105 1979 with Tamotsu Ootaki
"An Unstable Gene in *Phycomyces,*" *Genetics*, 92: 27–48.

106 1979 with M. Jayaram and D. Presti

"Light-induced Carotene Synthesis in *Phycomyces*," *Experim. Mycology*, 3:42–52.

107 1980 "A Valentine for N.I.H.," *Trends in Bioch. Sci. (TIBS)* (February):xii.

108 1980 with M. Jayaram and L. Leutwiler
"Light-induced Carotene Synthesis in Mutants of *Phycomyces* with Abnormal Phototropism," *Photochem. Photobiol.*, 32:241–245.

109 1980 "Was Bose-Einstein Statistics Arrived at by Serendipity?", *J. of Chem. Educ.*, 57:467–470.

110 1981 with M. K. Otto, M. Jayaram, and R. M. Hamilton
"Replacement of Riboflavin by an Analogue in the Blue-Light-Photoreceptor of *Phycomyces*," *Proc. Natl. Acad. Sci. U.S.A.*, 78:266–269.

111 1985 *Mind from Matter? An Essay on Evolutionary Epistemology*, ed. by Gunther S. Stent and Ernst Peter Fischer, Solomon W. Golomb, David Presti, Hansjakob Seiler. Oxford: Blackwell Scientific Publications.

in German *Wahrheit und Wirklichkeit: Über die Evolution des Erkennens*, trans. by Ernst Peter Fischer. Hamburg: Rasch und Röhrig Verlag.

Reviews by Max Delbrück

1943 *Arch. Biochem.*, 3:132–134; Review of *Virus Diseases*, by Members of the Rockefeller Institute for Medical Research: T. M. Rivers, *et al.* Ithaca, N.Y.: Cornell University Press, 1943.

1945 *Quart. Rev. Biol.*, 20: 370–372. Review of *What Is Life?*, by Erwin Schrödinger. Cambridge, Engl.: Cambridge University Press; New York: The Macmillan Company.

1946 *Rev. Sci. Instrum.*, 17: 133–134. Review of *Niels Bohr, an Essay*, by Leon Rosenfeld. Amsterdam: North Holland Publishing Company, 1945.

1968 *Eng. Sci.*, 31 (June):6.
Review of *Structural Chemistry and Molecular Biology*, ed. by N. Davidson and A. Rich. San Francisco: W. H. Freeman.

1970 *Eng. Sci.*, 33 (April):53–54. Review of *The Coming of the Golden Age: A View of the End of Progress*, by Gunther S. Stent. Garden City, N.Y.: Doubleday, Natural History Press.

Index